CO-ASP-852

Between East and West

⁓ Between East and West ⁓

Life on the Burma Road, the Tibet Highway, the Ho Chi Minh Trail, and in the United States

FU HUA CHEN, D. Sc., P.E.

UNIVERSITY PRESS OF COLORADO

© 1996 by the University Press of Colorado
Published by the University Press of Colorado
P.O. Box 849
Niwot, Colorado 80544
Tel.: (303) 530-5337

The University Press of Colorado is a cooperative publishing enterprise,
supported, in part, by Adams State College, Colorado State University, Fort
Lewis College, Mesa State College, Metropolitan State College of Denver,
University of Colorado, University of Northern Colorado, University of
Southern Colorado, and Western State College of Colorado.

The paper used in this publication meets the minimum requirements of the
American National Standard for Information Sciences—Permanence of Paper for Printed
Library Materials. ANSI Z39.48–1984.

Library of Congress Cataloging-in-Publication Data

Chen, F. H. (Fu Hua)
    Between East and West : life on the Burma Road, the Tibet Highway,
the Ho Chi Minh Trail, and in the United States / Fu Hua Chen.
        p.   cm.
    Includes index.
    ISBN 0-87081-434-6
    1. Chen, F. H. (Fu Hua)   2. Civil engineers— United States—Biography.
3. Civil engineering—China.   I. Title.
TA140.C47C47 1996
624'.092—dc20
    [B]   96-2732
                                                     CIP

10   9   8   7   6   5   4   3   2   1

紀念魏懷伯父

時年八十四

一九九四年於丹佛

陳學莘著

This volume is dedicated to
Uncle Wei Huai
to whom I owe my education and philosophy of life

Contents

Foreword

This is a moving story of loyalty, devotion, ingenuity, and ambition. It is the fascinating memoir of an adventurous civil engineer becoming a successful businessman and a suave politician.

Fu Hua Chen started his career as a civil engineer. He was given a "Mission Impossible," to construct the Burma Road and the Tibet Highway during World War II. Only through bold, imaginative, and unconventional approaches was he able to accomplish the mission.

This book is filled with sweat and tears, as well as pathos and excitement. It is written with great humor. The reader is carried away by Fu Hua Chen's life-and-death encounter with the Japanese invader in China and precarious experiences in jungles and highlands.

All this is but a prelude to the incredible scenario of how Fu Hua Chen built an "empire" of consulting engineering single-handedly from a garage shop in Denver, Colorado. His determination, ambition, and ability led him step-by-step to the top of the ladder. He won distinction in the academic and political arenas of the United States.

Fu Hua has performed a miracle. He has spanned the East and the West and deserves more credit for bridging the two than the medals and honors conferred on him can convey.

A biography, like a home movie, can be boring, but this life is stimulating, inspiring, and entertaining. I congratulate my brother Fu Hua for his outstanding and highly commendable account.

Ambassador Tai-chu Chen, Ph.D.

Civil engineers around the world have designed and constructed projects for the benefit of people throughout many centuries, but only a few civil engineers have contributed to two nations (China and the United States) as much as has Fu Hua Chen. His autobiography is written as an existing adventure, accomplishing engineering feats considered impossible in China.

Having received medals from the presidents of China and the United States, Fu Hua Chen significantly contributed to preventing the Japanese from conquering China in World War II. As chief engineer leading the construction of the Burma Road, he provided supply lines to defend against the Japanese invasion of China. Although he was captured, he was the only survivor of his engineering command and heroically escaped the Japanese forces.

A graduate of the University of Michigan and the University of Illinois, Fu Hua Chen describes his life as a civil engineer; his education, research, testing laboratories, and teaching; and his role as a leader of people. This book is difficult to put down because of Fu Hua's continuous desire to "make a difference." Even today, after over sixty years of adventure in civil engineering, Fu Hua Chen is leading research on expansive soils and is advising projects in China and the United States.

After moving to Hong Kong and starting a new career with the Hong Kong Public Works Department, Fu Hua Chen excelled in developing a material testing laboratory. Upon declining an offer to return to China, he migrated to the United States in 1957 and quickly established a reputation as an outstanding geotechnical engineer.

His civil engineering work involves hand-built roads (hundreds of kilometers in length), bridges, and dams. His life is not only about technical achievements but is also about how he managed thousands of workers and motivated people to accomplish the impossible at a level of personal sacrifice rarely seen. His daily life in China is sparked with humorous incidents and descriptions of the history and culture along the highway routes.

The civil engineering profession and the world as a whole are proud of Fu Hua Chen. This story unfolds with existing pride and respect. It will be noted in history as a one-of-a-kind example of how a civil engineer contributed to the enhancement of the quality of life for all people.

James W. Poirot
President (1993–1994)
American Society of Civil Engineers

Seldom, if ever, will one see Eastern and Western cultures interwoven in an exciting fabric of engineering accomplishment as is depicted in this book by Fu Hua Chen. This is the moving chronicle of an outstanding engineer and businessman's lifetime of understanding the history, culture, and people in two great countries — China and the United States. Vivid written pictures describe the completion of nearly impossible engineering tasks under extremely difficult conditions.

This is a book for everyone to read. Each person will find it inspirational. The reader will be enlightened and motivated and will appreciate more fully what life has to offer.

Fu Hua Chen's engineering achievements in China are awe inspiring. More important to him, perhaps, is his success in establishing a consulting engineering firm in the United States. Thirty years ago it was extremely difficult for a Chinese national to study and to establish an engineering business in the United States. Indeed, few were successful. It is even more remarkable that Fu Hua Chen could engender the trust of the professional community when the cultural environment within the United States was adverse at times. He was honest, ethical, and talented in both technical and business aspects of his profession, and he maintained strict adherence to the highest standards of quality.

We must give much credit to Edna Chen, Fu Hua's gracious and unselfish wife, for her role in his success. Without her unfailing support it would have been extremely difficult, if not impossible, for him to accomplish so much.

Fu Hua Chen — student, teacher, scholar, humanitarian, benefactor, patriot, engineer, entrepreneur, author, historian, philosopher, son, brother, husband, father, colleague. Knowing him is a life experience. Reading this book is next best.

Paul E. Bartlett, Dean Emeritus
College of Engineering
University of Colorado at Denver

Preface

My name is Chen Fu Hua. Chen is my family name, and Fu Hua is my given name. Contrary to the Western tradition, the Chinese place their family name first, followed by their given name. To avoid confusion I put Fu Hua first followed by Chen. Thus I am known here as Fu Hua Chen. Chinese do not have middle names. I still frequently receive letters addressing me as "Dear Hua" instead of "Dear Fu Hua." Maybe it is easier to adopt a Western given name such as Peter or James, but I am happy with my friends addressing me as Mr. Chen or just Chen.

Autobiographies are usually written by prominent figures such as past presidents, famous generals, popular entertainers, or notorious criminals. The lives of engineers are usually so dull that no one would be interested in reading about them. I have yet to locate an autobiography about the structural engineer who designed the Brooklyn Bridge or that of the civil engineer responsible for the Hoover Dam. I am among the millions of engineers who led a dull life, and my name will be forgotten long before I die.

I have several motives for writing this book. One is to offer readers a better understanding of the difference between Chinese and American culture. Another is to provide a historical account of the challenges faced by Chinese engineers fifty years ago. Also, I feel deeply indebted to the Unites States for allowing an immigrant like me to establish a career in my own private business. From my oriental point of view, I may disagree with many aspects of U.S. government, yet I have to admit that this is a land of opportunity, a land of freedom, and a land where one can fulfill one's dream.

I want my American friends to know that I am grateful for their advice and help over the past forty years. I am indebted in particular to Dr. Dennis J. DeSart for his preliminary editing and advice. Without his encouragement, this book could never have been finished.

 *Between East and West*

– *Prologue*

The year was 1942. It could have been May. In the warm weather, I wore only a shirt and light, khaki pants. I stood at the edge of a ravine more than fifty feet deep. Tropical vines and trees grew far below, but here at the top of the ravine the land lay completely barren. I had come here in the company of about twenty other engineers. We were the last to leave the western arm of the Burma Road, which stretched before us near the ravine.

My engineering corps and Chinese workers had built the Burma Road from Kunming, China, westward into Burma, which the Japanese occupied. Japanese troops that occupied northern Burma were advancing. This meant all the civilians who could flee were moving east into China.

Fifteen hours earlier I had joined my fellow engineers at a construction shed about twenty miles from this ravine. It had been a noisy night, and none of us could sleep. The bumper-to-bumper traffic of refugees had rolled on all night, and we had heard rifle and cannon shots in the distance.

The evacuation had started more than three days earlier. Convoys loaded with all kinds of belongings headed for Kunming, the capital of Yunnan Province. The fortunate convoys with better vehicles stayed on the road. In those that were less lucky, drivers had been shot by stray bullets, and vehicles lay overturned on sharp curves.

To keep the traffic moving, Nationalist (Guomindang) Chinese soldiers pushed the dead with their vehicles into the roadside ravine. There was no law. The strongest and the fittest survived in the scramble to leave.

As chief engineer I was in charge of my group, but we had no instructions from the Nationalist army with which we worked. Earlier I had left my engineers to find my way through the mountainous terrain to Lachu, the border town between China and Burma where the high military post was located. I had asked to see the commander in chief, General Yu Feipeng.

I waited patiently for hours. Finally the general's aide told me that the general would see me. I asked the general rather timidly when our small engineering unit would be allowed to evacuate.

The general, without lifting his head, bellowed, "You wait for orders!" He dismissed me with no chance for me to explain our uncertain existence on the road. I started to turn around for the door, then remembered that visitors were not allowed to turn their backs to a Chinese general. As I had been instructed, I walked backward to the door, still facing him. This incident had occurred almost a week before we stood at the edge of the ravine.

In fact, the general's bullying had been a pose. I later heard that in the middle of that same night he left his headquarters quickly wearing only his nightclothes and slippers, not even taking time to put on his boots. So then I understood why he ignored the group of engineers stranded at the top of the hill, eagerly waiting for his instructions: he was obviously in a hurry to leave himself.

I had returned to my engineers at the edge of the ravine, who were looking at the continuous line of traffic. It passed over the newly paved blacktop we had constructed during the past six months. I was heartbroken to imagine that the Burma Road, more or less my creation, might be destroyed by the approaching Japanese.

As I watched the refugees, I recalled all the hardships I had endured during the construction of the road. Now I had to leave everything behind and see to the safety of my wife and my three-month-old daughter. My wife came from a wealthy family and was not used to hardships. She had borne the privations of the last year in the rugged mountains at the western end of the Burma Road wonderfully well. Our house was merely a tool shed that had been built for the maintenance crew. She would look for me in the evening as I came down the hairpin curves of the mountain road. When she spotted my pickup, she knew I would be home in half an hour, and she would have our dinner prepared. She cooked on a charcoal stove made from a five-gallon kerosene can. Every day a local country girl about ten years old would come to help my wife with chores. The girl worked for only a good meal at the end of the day.

Around four A.M. the gunshots came closer and closer to our position by the ravine. Even without orders or instructions, I decided we had to join the long convoy leaving the area. All of my engineers began the evacuation in an orderly manner except for one. In his excitement, he started the pickup truck and headed for the road without us. I was forced to stop him with my pistol.

By six A.M. we joined the stream of refugees on the way to the bridge over the Salween River, realizing that if we could reach the bridge we might be safe. The distance between our station and the bridge could not have been more than twenty-five miles, but we could only move at a speed of five miles per hour in the growing traffic jam. By midday, we were within sight of the bridge when all hell broke loose.

The Japanese had disguised a small force as refugees. They were ahead of us, and they attacked and attempted to take command of the bridge. When we heard a major explosion, we knew the bridge was gone.

I learned later that the Chinese brigade guarding the bridge blew it up before the Japanese were able to capture it. In the process, the entire brigade of Chinese defenders was killed in the explosion or drowned in the river. To my knowledge, this act of heroism has never been recorded in the history of the Sino-Japanese War. The men in the brigade had to have known they would die by destroying the bridge, but they did it anyway.

Everyone on the crowded road scrambled in all directions, like a colony of bees after their hive has been struck. My engineers and I abandoned the truck and escaped into the ravine. Knowing the topography of the country, I realized that those of us still west of the river had only two alternatives: we could swim across the river or walk upstream to the town of Bamo, about a hundred miles away.

Hiking upstream would have been easier, but I did not know if Bamo had been taken by the Japanese. If they had, we would be walking directly into the tiger's jaw. I decided to head for the river, even though we would have to hike overland. The other refugees might return and jam the road, and the Japanese infiltrators were still loose in the area. Besides, with the bridge gone, the road merely led to the riverbank.

In the confusion, I had already lost most of my fellow engineers. A small group of us found our way east. Late in the day, we arrived at a deep ravine, still in the company of many other refugees. I had no idea where the ravine led, but with any luck it would lead to the river. We had no way of knowing until daylight.

We spent the night at the bottom of the ravine, something I would never forget. Huddled in constant fear, we could hear the heavy boots of the Japanese troops at the top of the ravine and an occasional rifle shot. We were powerless. All we could do was hold our breath when we thought someone came near.

If the Japanese discovered us with their high-powered searchlights and machine guns, none of us expected we would live. I was hungry and cold, as we all were. I had not had anything to drink since that morning. The hundreds of us trapped in the ravine shared the same agony throughout the night.

We were only a few miles from the Salween River Bridge, a suspension bridge with about a three-hundred-foot span. The bridge construction had begun in 1938, and it turned a new page in the history of bridge building. The main cable had to be imported from Rangoon, Burma. Since no road existed on the route, the cable, packed in huge drums, had to be rolled by hand over rugged terrain by laborers. Inch by inch and foot by foot, the drums were rolled through mountain paths and across streams toward the bridge site. It was a heroic effort — it took five hundred men six months to make the journey.

At the Salween, the cable had to be stretched across the river using a kite. First a kite was flown across, then a bigger rope could be pulled over. Finally, the cable was pulled across the river. The bridge was completed in 1940.

During that long night in the ravine, I thought about the struggle for this route. The Japanese occupation of Burma divided the Nationalist Chinese troops in southwestern China and their British allies in eastern India. If the Japanese could be driven out and the Burma Road completed from Burma to China, then supplies from India could be sent to the Chinese troops that were resisting the Japanese in southern and eastern China.

I had joined the work on the Burma Road in early 1941, four years after Japan had invaded China, which initiated Japanese involvement in World War II. After Pearl Harbor, the Japanese quickly attempted to cut off the Burma Road. The bridge over the Salween was the most vulnerable target, and the Japanese air force bombed the bridge repeatedly.

The bombs hit the deck of the bridge but missed the abutments, the tower, and the cables. Most of the bombs landed in the stream. The detonations left hundreds of stunned fish floating in the water after each raid. When we were working nearby, we had always celebrated the safety of the bridge with a fish feast after each bombing attack. The bomb damage was always repaired quickly by engineering crews stationed at the bridge for this purpose. The Salween, also known as the Nujiang or the Fury River, is one of the few main streams that flows north to south in China.

In Chinese history, the river and the area along the Burma Road were first mentioned around A.D. 250 in the Three Kingdoms period. In the classic historical novel *The Romance of the Three Kingdoms,* the legendary statesman and strategist Zhuge Liang led his army into Yunnan Province and attempted to cross the Fury River, known as the Lu at that time. His mission was to conquer the rebellious native clans in the region. According to the novel, "In May he crossed the Lu and struck deep into Bamo country." Stories were told about the river valley being clouded by an evil mist, which caused outsiders to become ill and die. Today we know that the mist is generated by the dense tropical growth, which is infested by mosquitoes. The sickness is acute malaria.

In fact, this disease had posed a great problem for the builders of the Burma Road because it could be deadly. More than half of the labor force was infected at one time or another. The clinic passed out tons of quinine with little effect. We heard that a new remedy called atabrine, recently produced by a German laboratory, was highly effective against malaria. However, in China this drug could only be obtained in Shanghai, which was held by the Japanese. We sent special agents into Shanghai, and they were able to smuggle out a small amount of this new wonder drug. I was fortunate enough to use some of it to treat my illness.

Back in the time of Zhuge Liang, legend said the deadly mist could be conquered only by praying to the river god and sacrificing one hundred human skulls to appease the god. Legend also declared that the river god would not allow a human being to swim the Fury River without the sacrifice of one hundred people.

Workers were lost to disease, accidents, and war. However, we named the bridge the Benevolent. I later named my daughter after the bridge.

At the earliest light, we began to walk. By dawn we were out of the ravine — and, to our horror, we marched straight into the Japanese garrison. I was staring into the muzzle of a soldier's rifle.

I would never forget his face. It was expressionless except for the slightly curling lips, which seemed to me to be extremely cruel. Today, after more than fifty years have passed, I still see the soldier's face in my dreams. We all knew the Japanese did not keep prisoners here. I heard the rattle of a machine gun nearby, and I knew that the slaughter of prisoners had started.

In a split second, I experienced the famous phenomenon of having my entire life flash before me in minute detail. I had never really believed the stories about this before. Now, looking down the barrel of a gun, I personally experienced it.

The Chinese believe this represents the judgment period. It provides a review of one's life as recorded by some supernatural power. Legend says that if you have done nothing but good all your life, at that moment you will feel warm and calm, as if you are bathing in the sunlight. On the other hand, if you have done evil deeds all your life, you will feel cold and irritated and yet hot, as though burning in fire.

I felt neither warm nor cold but numb, perhaps because I hadn't lived long enough. I was only twenty-nine years old.

Bridge over the Salween

Bumper-to-bumper traffic

*First fleet of convoy on
the Burma Road*

Partially completed Burma Road

– Chapter One
Century of Humiliation

China suffered continuously during the later years of the Manzhou (Manchu) Dynasty in the nineteenth century, the early years of the Republic of China in the early twentieth century, and the Sino-Japanese War, which was later assimilated into World War II. Very little occurred that benefited China during this period, during which I spent my formative years.

For centuries, the Chinese considered China to be the center of the earth and believed all other countries were barbaric. This myth was shattered in the early nineteenth century when foreign warships steamed along the coastal cities and up the Yangzie (Yangtze) River. The national humiliation continued during the reign of the dowager Empress Ci Xi. The British, Germans, French, Italians, Russians, and Japanese wanted to claim portions of China. Suddenly, the Chinese developed a deep fear of the "foreign devils." They felt the westerners were doing everything right and that the Chinese could accomplish nothing.

In 1830 the British discovered a fortune in importing opium to China. The opium was grown in India, which was under British rule, and was shipped to China. The amount consumed in China was staggering. Opium consumption became a status symbol for the upper class and an incentive for the laborers.

While their opium trade thrived in China, the British realized the danger of smoking opium and enacted laws with stiff penalties for opium users in Britain. The opium trade was a powerful tool for bringing cash to the British Empire and weakening the Chinese people. Many historians consider the era to be a serious blot on the honor of the British.

One outstanding Chinese official named Lin Zexu, governor of Guangzhou, stood up to this drug trade. In 1839 Lin ordered that a British opium shipment be burned. He also appealed to the emperor to enact laws that would ban the opium trade and punish the users. Alarmed at the possibility of losing their extremely lucrative trade, the British sent a fleet of warships to escort trading vessels bringing in the opium. Lin fought back, even when the empress ordered him to surrender, thus starting the Opium War.

The British quickly diverted a warship to the defenseless ports of Shamen and Dinghai near Shanghai and pushed on toward Nanjing. The Imperial government surrendered; Lin was dismissed and sent into exile. The island of Hong Kong, considered a barren wasteland, was ceded to Britain.

It is noteworthy that we in the United States are aware of the harmful effects of cigarette smoking, such as cancer, heart disease, and respiratory damage. Whereas we try to curb smoking and even ban passive smoking in this country, we have promoted the sale of tobacco products overseas. Sales by U.S. tobacco companies have tripled in recent years, and even our government encourages the export of tobacco. Could the story of the Opium War teach us something?

In subsequent years, European powers and Japan won further concessions in China's cities, coasts, and river valleys. By the late nineteenth century, these foreign nations virtually ruled many of China's most important locations. Perhaps the only reason China was not completely conquered was that none of the foreign powers would allow any one of the others to gain total control.

My father was born in 1880, only fourteen years after Sun Yixian (Sun Yat Sen), better known in China as the father of the Republic of China. By the early twentieth century my father had become one of Sun Yixian's faithful followers. My father had an intense hatred of the Manzhou and all those who served Manchurian masters, who the Chinese called their running dogs. He argued that for China to shed its years of humiliation, the country would need to be completely westernized. He set an example by mastering English.

In 1911 Sun supervised the overthrow of the Manzhou Dynasty. A year later, in the city of Nanjing, he became the first president of the Republic of China. His presidency was brief, lasting only forty-three days. Yuan Shikai, the general who had defeated the Manzhou armies, took power by force and declared himself the new emperor in 1915. He died in 1916, before he was enthroned, and the rule of China fell to regional and provincial warlords.

Sun's democratic loyalists, organized in the Guomindang political party, had new enemies. They fought first for the overthrow of Yuan and then against the warlords who divided China among themselves. One of the revolutionists' headquarters was located in the British Concession in Shanghai because they could not be openly persecuted there by Yuan's forces or by the provincial governor.

At that time my father worked for the customs services under the British administration in Shanghai, making use of his knowledge of English. We lived in a two-story bungalow where my mother raised her three children. My father spent most of his spare time alone in the workshop downstairs.

I was told my father originally named me Chen Sheng. Chen Sheng was a peasant who led a revolt at the end of the Qin Dynasty around 200 B.C. He was my father's hero, and they had the same last name. I was called Ah Sheng by my parents. After the death of my father, friends pointed out to my mother that Chen Sheng's revolt had failed, and he had been executed. They convinced her to change my name to Fu Hua, although she still affectionately addressed me as Ah Sheng.

My mother worked as a nurse. She had received a good deal of modern medical training, which was rare at that time in China. Also, she knew many of the missionaries who continued to bring Christianity into China.

The bitterest enemy of Sun Yixian's revolutionary movement in Shanghai was provincial Governor Xu Baoshan. He was a relentless fighter and a ruthless military officer; he beheaded many of Sun's followers when he caught them outside the protection of the British jurisdiction. He also had agents working secretly in the British Concession.

My father learned how to make bombs from *Popular Mechanics* magazine, from which he created homemade bombs and killed some of Governor Xu's agents. Xu himself was well guarded, and it was almost impossible to get near him with a bomb.

Then my father found out that Xu was a passionate collector of rare manuscripts. In China, old manuscripts were rarely bound but were held together with two wooden boards at the top and bottom of the stack. My father conceived the idea of hollowing out a thick manuscript and placing a handmade bomb inside set to detonate as soon as the top board was removed.

Gathering the bomb materials and an appropriate manuscript took many months. My father had to work cautiously because of the risk of accident; he also had to work in secret because the penalty for this conspiracy would be his death at the hands of Xu. One Sunday afternoon while my brothers, my sister, and I were upstairs with our mother, we heard a tremendous explosion from his downstairs workshop that shook the entire building. My father died instantly. He was only thirty-four years old.

Everything in the workshop was scattered and broken. Within minutes, British detectives arrived at our house. They took my mother to the garrison for questioning. Friends of my father's took care of the burial and cared for us children. The British put Mother in jail and questioned her about the names of the revolutionists. For three days and nights the British interrogators could not get a word from her. Finally, they released her. When we were reunited back at home, we saw her break down and cry for the first time.

A month later my brother was alone in the bedroom when mother walked in. "Who were you talking to?" she asked. "Why to daddy, of course," my brother answered. Children can sometimes see things adults cannot.

*My father
Chen Mu, the
revolutionist*

I have very little recollection of my father except for what my mother told me about him, and she said little. Once I saw a copy of *Aesop's Fables* on the bookshelf. The hard binding seemed to be covered with pinholes, and I asked Mother what they were. She took the book from me, and I never found it again. I later realized the book had been in father's workshop when he died and that the spots were bloodstains.

My mother, who had a Christian background, advocated peaceful solutions. She wanted us to remember Father as a peace-loving man rather than as a terrorist. On the day of my father's death, she placed plates of fruit in front of my father's large portrait. We assembled to pay our respects by bowing three times, and we performed this simple rite every year until my brother went to college. Even after my father died for the revolution, Guomindang loyalists held meetings at our house. Warlords still ruled, but the Guomindang had not given up. I got to know them well.

– Chapter Two
From Beijing to the United States

I attended high school in Beijing, having moved from Shanghai. The school had been founded by missionaries from London. Their goal, of course, was to bring the word of Jesus to the nonbelievers. As part of the price of education, all students were required to participate in the morning prayer meetings and Bible study sessions. The missionaries, to their credit, were extremely dedicated. History, philosophy, and geography were taught in English. I benefited from these lessons, and acquired a knowledge of basic English.

The school's administrative philosophy was based on the English tradition, which allowed no nonsense from students. Discipline was strictly enforced. Any slight violation of the rules brought corporal punishment down on the offending student. The principal was a tennis player, and he would use a yardstick with his service stroke to hit the student on the open palm. If the student flinched, the stroke was considered a fault, and the blow was repeated.

No students in my class escaped such punishment. To add insult to injury, I did not dare to go home and tell my mother of the punishment. If I did, I would likely be punished at home, too. I find it amusing that today teachers can be sued for such behavior and possibly go to jail or lose their jobs. Yet I am grateful now for the rigid discipline, and I think it has probably benefited me throughout my life.

Fifty years later, I visited my old high school in Beijing. The school building and the church were still standing, now under Communist control. The school gave me quite a welcome as a returning alumnus. I learned that only a week before, the school had welcomed another former graduate named Yang Zhenning, a Nobel prize winner in physics. Maybe there was merit in the strict discipline in our early years.

Coeducation was not practiced in China when I was a student. The sexes were strictly segregated. Of course, we talked and joked about sex all the time.

13

Our favorite topic for teasing, though, was usually homosexuality. When an older boy befriended a younger student, we all pointed fingers and labeled them homosexuals. In fact, in China homosexuality was widely practiced among men, especially in the upper class. It did not carry the serious stigma it does in the West. Homosexuality was considered one among many natural practices, and questions were not raised about the morality of such acts. Homosexuals did not lose social standing, and they never openly discussed their sexual behavior, just as a man and a woman would not openly discuss their sex lives in China at that time.

The traditional attitude in China toward homosexuality has a long history. In the year 6 B.C., Emperor Ai Di dearly loved a handsome young boy. One day the emperor was lying on his bed with the young boy beside him asleep on the sleeve of his magnificent silk robe. Forced to get up, the emperor preferred to cut off his sleeve rather than disturb the sleeping boy. Ever since, the term *cut sleeve* in Chinese has come to mean homosexual. Only in the Christian world does homosexuality become a big issue. Politicians and lawyers, along with groups such as the Moral Majority, fueled the debate until it became a national battleground. They don't realize that as soon as we ignore such issues, they will evaporate. There are more urgent matters to be resolved. Americans seem to believe that all social and moral problems can be addressed by legislation and then litigation. We have more attorneys than are found in the rest of the world, and they have created more problems than they have helped to resolve.

One major difference between the Chinese and the U.S. educational systems is the high regard Chinese students have for teachers. The ancient traditions of Chinese teachings demand that the status of teachers is the same as that of parents. Historically, Confucius was revered as a great teacher, and the role of a teacher remains widely respected. The teacher's words are law. In the classroom, discipline is considered primary. No nonsense from the students is tolerated. Parents and teachers work hand in hand toward the same goal of bringing up a well-educated student.

The high school curricula illustrate this difference. In China we finished physics, chemistry, and biology before graduation and started calculus and analytical geometry in preparation for our freshman year in college.

It seems to me that the schools in the United States are filled with students who have no respect for the teachers. The mood is one of equality between students and teachers. The teachers can be sued by the students for abuse. At the same time, parents believe that as long as the students are in school, it is the responsibility of the teachers to train and educate their children. This conveniently transfers the parents' responsibility to the teachers and the schools. The public believes the way to improve education is to spend more money on it. Have we seen any improvement in the last ten years, when the funding for education has tripled in many states?

On September 18, 1931, the Japanese seized Manchuria and set up the puppet regime of Manchukuo. I was enrolled as an engineering student at the University of Nankai, known as the cradle of the student movement in China, at the time. Zhou Enlai, a graduate of Nankai, actually founded the Communist Party in Paris several years before Mao Zedong became prominent. Zhou was to become a beloved figure among the Chinese. He had been insulted by John Foster Dulles and could not forget it. Dulles had refused to shake his outstretched hand at a public occasion. His skepticism about Americans was rooted in part in this incident.

The cloud was lifted somewhat when President Nixon offered to help Zhou with his coat in Beijing in 1972. Nixon's gesture considerably thawed the feelings of the Chinese toward the United States. Sometimes seemingly insignificant acts mean a great deal in the minds of the Chinese. Since this act was not political rhetoric but a gesture from one person to another, it was likely to be seen as more significant than all the diplomatic correspondence between Beijing and Washington.

During the tumultuous years of the "cultural revolution," Zhou saved many Chinese from persecution, and without Zhou's intervention the outcome of the cultural revolution could have been much worse than it was. The University of Nankai has a large monument in honor of its distinguished graduate.

In 1932 thousands of students from Nankai and other universities in Beijing went to Nanjing to appeal to Jiang Jieshi (Chiang Kaishek) to get tough with the Japanese. However, I saw that the future of China was filled with social chaos and war. I had little to gain from continuing at the university.

I applied to the University of Michigan and was accepted. Since I was the son of a Guomindang martyr, the government would pay part of my expenses. I went to the United States, the land of golden opportunity, to help China fulfill the dreams of Sun Yixian.

In Chinese, the nineteenth-century term for the United States was *Jiu Jin Shan,* or Old Mountain of Gold; now the term refers more specifically to San Francisco. Originally a reference to the California Gold Rush, it continued to imply a nation of open opportunity. Upon my arrival, I witnessed for the first time the true status of the Chinese immigrants in the United States.

Most of them had come from Guangdong Province in search of the golden bricks with which the streets were paved in San Francisco. Many were veterans of World War I. A large number were illiterate, so occupational doors were closed to them. Those who had sent money home to China or returned wealthy had not told us about the ones who failed to do well.

The early laborers searched for gold but also did the jobs that were available as migrant workers and domestic servants. Often, they opened their own restaurants and laundries. In San Francisco and smaller cities and towns, they created Chinatowns.

In the late 1860s, the Central Pacific Railroad hired thousands of Chinese workers to do the backbreaking work of building the railroad across the Sierra Nevada. They cooked their own meals and lived in their own tents. Methane gas made the tunnel work especially dangerous, and when the local workers refused to go in, the Chinese did so. In summer 1866, the Chinese were lowered down a cliff face in baskets, where they drilled holes and placed and lit dynamite. Many were killed when they were not pulled up quickly enough or when the ropes holding their baskets snapped. They also faced considerable racial prejudice during these years, when the expression "not a Chinaman's chance" originated.

After years of tremendous toil, in 1869 the railroad that joined one coast of the nation to the other was completed. The huge celebration in Utah excluded the Chinese, and throughout the ceremony just one brief mention was made of their labors. A hundred and twenty years later when the Union Pacific remembered the event, again nothing was mentioned about the Chinese laborers.

The anti-Chinese movement gained momentum in the 1880s. One of the worst atrocities occurred at the Rock Springs massacre in Wyoming. In September 1885, a crowd of white miners gathered and drove all of the Chinese immigrants out of the mines and back to the Rock Springs Chinatown. Later in the day, an angry mob of roughly a hundred and fifty white people headed for Chinatown. Shots were fired, and the Chinese fled to the hills.

"The Chinamen were fleeing like a herd of hunted antelopes," one observer wrote. Volley upon volley was fired at the fugitives. Those who were not quick enough were shot or perished in the flames that destroyed Chinatown. In all, twenty-eight Chinese were killed and fifteen were wounded. The Rock Springs massacre generated protest on the East Coast. The U.S. government paid $150 thousand in indemnities to the families of the victims, but none of the white miners was indicted.

By the time I arrived in 1933, chop suey and laundries had been the chief U.S. symbols of the Chinese for several generations. To help meet my expenses, I worked in a Chinese restaurant in Ann Arbor each evening washing dishes. Automatic dishwashers did not exist at that time, and I had to dip my hands into the hot water for hours until they were raw, all in exchange for an evening meal.

It was not easy for me to swallow my humiliation when I was followed by children dancing and singing "Chin Chin Chinaman" or when I looked at the sign in the window of a restaurant that said "We do not serve Chinamen." It was not easy to tolerate the way movies and comic strips characterized my people by showing Anna May Wong or coolies always holding a long knife and wearing a flying queue; there were similar images in *Terry and the Pirates.* I could not find a rooming house on my college campus in Michigan that would accept me, and at that time the term *civil rights* did not include housing. As a student at the University of Michigan, I swallowed my humiliation silently.

Everybody in Chinese restaurants, shops, and laundries throughout the United States worked very hard. The laundry workers and chop suey restaurant owners worked sixteen hours a day, and many managed to save money. They used their savings to send their children to college to become professionals — teachers, doctors, engineers. With hindsight, metaphorically, perhaps the streets *were* paved with gold. Since the time when I moved to Michigan, the general U.S. image of the Chinese has gradually changed.

At the University of Michigan, my first experience with racial discrimination occurred when the football team debated whether to play against a southern team with black players. A rally was held at the student union, and thousands attended. That was the first time I went to a podium to speak publicly, and I argued in favor of playing the game. My viewpoint prevailed.

Michigan had a powerful football team, with Gerald Ford as the captain. Later that season, when Michigan beat Ohio State in Columbus, Ann Arbor went crazy with celebration. More than forty years later, I talked with President Ford at his Vail home, and he still remembered the celebrations distinctly.

I believe prejudice is inherent in human nature. It is not a vice or a crime to have prejudices. In China, prejudice is common not only against foreigners, who Chinese refer to as foreign devils, but Chinese also harbor prejudices against one another. Northern Chinese are prejudiced against southerners. In Shanghai, which has an area of several square miles, those who live south of the river look down on those who live north of the river. As communication improves and contact among people becomes more frequent, prejudice gradually diminishes. Enacting legislation cannot eliminate prejudice.

At the University of Michigan, the Chinese students decided to solve the housing and restaurant problem by organizing our own fraternity house. That was how Alpha Lambda got started. I was one of the founding members. Our fraternity had strict academic standards and would only admit students with solid academic records. We provided our own housing and cooking. One of our distinguished members was Wang An, founder of the electronic giant Wang Laboratories. Alpha Lambda is still a strong organization on U.S. campuses and in Hong Kong and Taiwan.

In 1964 Lyndon Johnson declared the "Great Society" in an attempt to eradicate poverty and the problems of minority groups. Johnson's programs have become so costly that they have nearly bankrupted the U.S. economy, yet the social and economic ills they were supposed to address are still with us.

I think there is something to the value of labor. In 1644, when the Manzhou conquered the Han under Emperor Qian Long, the empire flourished. The Han people realized they could not defeat the Manzhou militarily, so a scholar suggested to the emperor that the Manzhous are a superior race, and the Hans are born to be slaves. It is not proper for the superior race to engage in manual labor or minor work. Its members should be entitled to a pension the day they are born, and this should continue throughout their lives. The Han, as an inferior race, should spend their lives doing the work of slaves and serve the master race. The emperor was delighted at this flattery and ordered that the day a Manzhou was born, he or she would receive a pension and would never have to do any work. In a mere two hundred years, the Manzhou were so weakened that they lost the spirit and the strength of their ancestors. Today, as an independent people, the Manzhou have virtually disappeared. A similar event can occur again today, as people lose the pride of accomplishment.

When I finished my B.S. degree in civil engineering at Michigan, I went to the University of Illinois for graduate work. I spent my time in serious study, and I also began to wonder whether my engineering education would be of any value to China. My roommate was a brilliant student, and we discussed the possibility of applying to West Point. I was denied admittance, but after a long and agonizing wait, he was accepted, aided by a letter from a senator. It saddened me to learn later that my roommate was killed in the first encounter of the Sino-Japanese War at Shanghai.

At the University of Illinois I experimented with a mixture of cement and soil. My co-worker was Joseph Linderbrand. We worked well together in the soil lab, testing different substances and combinations, but the janitor complained bitterly about having to clean up after us. Thirty years later, I called on Mr. Linderbrand in Chicago after he had become the head of the Portland Cement Association, and we both remembered how that janitor had objected to the mess we made in the lab.

That was one of my happiest periods in the United States. After I got my degree from Illinois, I decided to spend a year in Europe and see that part of the world. I stayed for a time at the Imperial College in London, but my only accomplishment of note was to lead the college table tennis team to a win against University College of London.

During my European tour I attended the Berlin Olympics in 1936, where Jesse Owens won his gold medals. Hitler's domination of Germany was at its peak. I was fascinated by the way Hitler charmed the German people. At the 1936 Olympics, Hitler was greatly annoyed by the performance of Owens, a member of a race Hitler considered to be inferior.

The Germans were pleased with the performance of the Japanese. I had an embarrassing moment when some Germans mistook me for a Japanese and saluted me, raising their hands and saying "Erste Japan." However, I remained silent, thinking that one day I would see China surpass both Japan and Germany in the Olympics.

During both my first years in the United States and my trip to Europe, I observed cultures other than my own. Historically, the Chinese had been able to accomplish a great deal in both science and technology, but they had difficulty cooperating instead of competing with each other. The strong cultural identity of the Chinese had not always extended to a loyalty toward national cooperation. Both the United States and Japan understood the value of cooperation within their own countries, and both had become major industrial powers. To compete in the modern world, the Chinese had to learn to cooperate within their own society.

Early in 1937, when I had completed my education, I returned to China to make my contribution.

— Chapter Three
The Xian Highway

I had little idea where best to use my engineering talents in China, but I knew that building roads would help any developing nation. So I thought I could use my knowledge from the U.S. universities I had attended to build four-lane highways in China. I traveled to Nanjing, then the capital of the Guomindang government under Jiang Jieshi and tried to find work.

I soon realized that what I had learned in school was far different from what was needed to be a highway engineer in China, where an engineer needed the strategizing ability of a general, the eloquence of a diplomat, the courage of an explorer, and the ingenuity of an inventor. Perhaps most important, one had to withstand severe physical hardships.

As I learned my way around the government bureaucracy, I discovered that engineering opportunities did exist. The Highway Administration in Nanjing wanted to build an all-weather road from Xian to Lanzhou without using concrete or asphalt. I suggested building the road with soil stabilization, using what I had learned while angering the janitor at the University of Illinois. Soon I had a job as a junior engineer on the Xian project.

The city of Xian had a long, glorious history in China. Long before the birth of Christ, the first emperor of China, Qin Shi Huang Di, consolidated his conquests and completed the Great Wall. He chose Xian as the capital of his vast empire. He built a magnificent palace surrounded by nine rippling streams that all led to the Yellow River, irrigating the land outside the city. Qin collected the finest art in his empire, but it was all destroyed when the empire crumbled a short time after his death. According to historians, the fires in Xian raged for more than three months and reduced the timber and masonry of the palace to ashes.

In 1974 three shallow vaults were found buried in modern Xian that contained several thousand terra-cotta warriors and horses. A museum, which is a major tourist attraction, was built on the site. Archaeologists now know the location of the emperor's tomb, but the bulk of the excavation has yet to be done.

The mausoleum is actually a lavish, gigantic underground palace. According to historical records, it was roughly 140 feet high with a three-mile circumference. Dug deep into the ground, it was sealed from groundwater with molten bronze. The tomb contained entire courts, with places reserved for officials, and a great horde of treasure. Quicksilver was pooled to create seas and waterways with mountains of jade.

Today a mound of tamped earth 160 feet high still stands above the mausoleum on a base about 1,600 feet square. The mound is landscaped on four sides like a pyramid. The mausoleum was so solidly constructed that it defies grave diggers. Archaeologists believe the mausoleum has never been entered. The main excavation has been postponed until basic preparation work has been completed.

Xian, meaning "Western Peace," was the capital of many dynasties at various intervals during a period of over two thousand years — sometimes under the name Changan, meaning "Eternal Peace." The Tang Dynasty, considered the Golden Age of China, made Xian its capital for three centuries as Changan. At this time the city stood at the height of its power and prosperity. Particularly during the reign of emperors Tai Zhong and Xuan Zhong (627–741), the economy and culture of the Tang reached a new peak. Changan was the world's largest metropolis at that time, equal in importance to Baghdad and Constantinople and six times the size of present-day Xian.

Tombs of many emperors dotted the surroundings of Xian. The most prominent is that of Empress Wu Zetian, the first female ruler in China. Since 1949, archaeologists have undertaken large-scale excavations in Xian, bringing to light thousands of cultural objects that illuminate China's ancient civilization in a new way. Even today, farmers continue to find pieces of pottery, which they discard as worthless.

After a three-day trip by train, I arrived in Xian to build the road. Nothing but yellow loess, which is a dry, dusty soil deposited in the area by constant wind, remained outside the boundaries of Xian as it was in 1937. Through the centuries since emperors had ruled the Chinese empire, nature had altered the topography of the entire area. The nine streams, once the subject of lavish praise by poets, had disappeared. Vegetation could no longer survive in the severe weather where dry, northern winds blew out of the Gobi Desert; without the precious irrigation water, only a local desert of barren loess remained. Driven by drought and famine, many of the local people drifted southward to survive. Only the poorest and weakest remained to await their end.

I took a room in the Grand Hotel in Xian, an establishment that usually catered to dignitaries and foreigners. The following day I reported to the engineering office. Chief engineer Liu Rusong, also a University of Michigan alumnus, greeted me with enthusiasm until I told him where I was staying.

Liu frowned. "Engineers who work here need to be tough. By staying in such a luxurious hotel, you are showing that you cannot withstand the hardships that are a central part of life here."

Fortunately, I did not lose my job. Liu gave me a chance to prove myself. Together, we examined the route of the proposed highway.

The first stretch of the proposed road was totally devoid of rocks and gravel. Nothing lay along it but mile after mile of yellow loess. It was the dry season, and we had to wear goggles to protect our eyes from the blowing dust. Liu told me that during the rainy season the trail turned to mud, and driving became impossible. Traffic could not resume until the sun had returned long enough to dry out the surface of the loess.

To build an all-weather road, one normally needed a source of gravel. The nearest source was over fifty miles away, and we had no means of transport. Clearly, the cost of hauling the gravel would be prohibitive, and I was asked to find a way to stabilize the soil without using gravel.

I suggested my pet project from my university days, combining cement with soil. Chief engineer Liu hesitated at first. As I debated the alternatives, he finally agreed to build a test section on the road to see if the idea had any merit.

A French engineer acting in an advisory capacity opposed the plan. "Cement and soil do not go together," he remarked. "I have never seen it done, and I don't believe it will work in this part of the country."

I left Xian one bright summer day to go to another historic city, Xianyang, about twenty miles to the west. At Xianyang I found a cliff face composed of nothing but loess. Thousands of caves had been cut into it by hand. They had no windows, and the only furnishings were couches, tables, and benches, all made of loess. It amazed me that the natives were able to do so much with so little. Except for the muddy smell and lack of ventilation, the caves were fairly comfortable. People told me the caves were rather warm in the winter and cool during the hot summer. The couches were heated from an outside tunnel and stayed warm throughout the day and night. People probably spent most of the day on the couches, and they served as both the dining area and sleeping quarters.

As we prepared the engineering team and the construction crew, our first priority was to locate a decent hotel. Everywhere we went we were told they had no room. The management simply did not want us around. We soon discovered that all the hotel rooms were occupied by prostitutes, and the only way to get a room was to negotiate with one of the women. If we paid enough, they would find a place for us. Once we knew how the system worked, we got along fine.

In fact, the money was well spent, because our hotel had the best food in town. It became a home away from home, since most of us were single. We wined and dined in the evenings, forgetting the trials and hardships of the day.

I later learned that the women came from Suzhou and Yangzhou on the eastern coast of China. They had traveled thousands of miles, probably taking a year and working their way across the country a little at a time. At a given location, if they found no demands for their services, they would drift westward. They journeyed through deserts and crossed rivers during bitter winters and scorching summers. I found that some of them had been as far west as Xinjiang Province, north of Tibet near the Soviet border.

They had no hope of seeing their homes again, but they deserved a great deal of credit. In a sense, they were true pioneers of China's west. No town large or small would exist without them. Although the politicians made a great noise about developing the northwest, none of them actually went there. These women were silent warriors who took part in the development of the region and received no notice for their contribution.

We completed the test section of the road. The result seemed satisfactory, although we did not have time to observe its use for an extended period. Liu, a decisive man, decided to go ahead with the work on the rest of the route.

We needed a laboratory to carry out the soil-cement stabilization. I built one in the field with whatever I had at hand. A frying pan served as a beaker; a kerosene stove acted as a Bunsen burner. I used a butcher's scale for a balance and soup spoons to stir and mix. The local people thought I was cooking when they saw me setting up in the field.

To compact our stew of soil and cement, we needed a large, heavy roller. Our roller was cut from solid rock into a huge cylinder. It weighed about five tons, and fifty men were needed to pull it.

Water was so rare in this area that local people used it only for drinking. The natives seldom washed even their hands, and a bath was a luxury beyond imagination. The residents had a saying that one only bathed three times in one's life: once at birth, once before one's wedding, and finally before burial.

We used empty fifty-gallon gasoline drums to carry water for our construction needs. The drums were loaded on horse carts and were hauled several miles to the construction site. Because of the scarcity of water, we had to be careful with it.

Whenever we returned to our hotel after a long, dusty day at work on the loess, we asked for water to clean our faces. As distinguished guests, we were given a bowl the size of a soup bowl in which the water was dark brown until the silt settled. Then we used the water to mop dust off our faces.

The lack of water in northwestern China resulted from centuries of neglect. Although the region had been forested at one time, the trees had been cut down ruthlessly without replacement, and the soil could not hold rainwater. The level of the groundwater had gradually receded to more than sixty feet below the surface. The land could become fertile once again through planting, water diversion, and irrigation, but in the late 1930s the government had few resources to invest in such a project.

The present regime has done a great deal to change the environment of the area through soil conservation and irrigation. Consequently, the lives of the people in the region have improved considerably. The people today enjoy clean water and sound health.

We got the soil-cement project underway with Portland cement, water, the stone roller, and some laboratory control, as well as a great deal of hand labor. After many months of hard work, to the great joy of everyone involved we completed ten miles of the road. My first professional engineering project was finished.

Unfortunately, my excitement did not last long. Several days after we opened the road to traffic, I could see that it was a failure. We had built it for motor vehicles with ordinary rubber tires. However, much of the traffic here consisted of horse carts with iron wheels. These narrow wheels simply chopped the surface of the road to pieces. We asked the local magistrate for help. He promised to take some sort of action, but he had too few police to patrol the road and stop the iron-wheeled carts from using it.

In the engineering sense, the soil-cement experiment did not fail because of poor design or inferior work. I had not foreseen carts with iron wheels using the road. Naturally, those who had been opposed to the project from the beginning sneered triumphantly, "What did we tell you?"

Liu cheered me up by saying, "This is your first job. The outcome is not as important as the fact that you gave it your best effort." That would have to do. Fortunately, I had dispelled any feelings of disrespect because I had initially stayed at the Grand Hotel. My fellow engineers gave me a nickname loosely translated from Chinese as "Toughie Chen."

During these years I learned a great deal about engineering from all the projects I did in China. We didn't have the luxury of trial and error, so a project had to work the first time. Several years later I tried my soil-cement idea again on a short stretch of road in the city of Chongqing (Chungking). This time, no animal-drawn, iron-wheeled carts were in use, and the road lasted for several years.

Chapter Four

A Road in Fourteen Days

Even before the soil-stabilization work had ended, Chief engineer Liu asked me to return to Xian for vital instructions. When I arrived, Liu showed me a telegram from Generalissimo Jiang Jieshi. He had explicit instructions for a new project: "It is of vital importance that the road from Xian to Baoji be in operational condition for military traffic by September 20, 1937. The magistrates of the various counties along the route will give you every possible assistance in procuring labor."

"This is the third of September," I said.

"Yes, so that gives us exactly seventeen days to do it," Liu answered. "Do you think such a project is possible?"

"I don't see how. The distance is about two hundred miles. In its present condition, it's no better than a rough trail used for pack animals."

"You have had no experience with the generalissimo's orders," Liu said patiently. "The question is not whether you can do it, but how soon you can get started."

"Am I supposed to handle this job?" I asked.

"What do you think I asked you here for? And I must warn you that an order from the generalissimo is no joke, and there is no room for argument," Liu replied.

That ended our conversation. I was given around fifty engineers, six radio transmitters, and an old Russian-built truck. Most important, Liu gave me duplicates of the generalissimo's order.

Given the schedule, I decided to start work immediately and not spend time studying the entire route. I made one last rough check of the map and got started. First, I assigned three experienced engineers to three sections of the road and told them to start on the detailed preparations for construction. They also had to choose a location for their respective headquarters so I could establish radio communication with them.

We got started on bicycles the following morning to inspect the existing trail and to negotiate with the various county magistrates for the necessary laborers. I had done a great deal of cycling in high school, and at the University of Michigan I had exercised on my bicycle every day to build up my leg muscles for ice hockey. Now I realized that the bicycle was easily the fastest and most reliable means of transportation for highway engineers in country like this.

Even so, long-distance cycling on this route was arduous. The trail was rough and had many steep grades. We had to carry our bedding and folding cots, which weighed about thirty pounds. In addition, taking these bicycles over the mountainous terrain was rather different from the long-distance cycle races that took place in France and the United States because the bicycles had no gears. It was necessary to push the cycles up the hills rather than pedal. Going downhill was a completely different matter; since we had no spare parts to replace the brakes, we pressed the soles of our shoes against the back of the front wheels to brake. Later, to make sure our shoes could stand the friction, we equipped them with heavy rubber cut from used tires.

Along the road, especially going uphill, travelers commonly encountered road bandits. Merchants who carried a few commodities on their bicycles were easy prey for these bandits. Since the bandits knew we were road builders and that some of us held high ranks in secret societies, we were left alone. Some of the secret societies had political affiliations, and others had a reputation for engaging in organized criminal activity.

By noon the first day, after three hours of continuous cycling, we had covered about twenty miles and reached the site of a small county office. I presented my card to the guard, who read it carefully and then showed me to the reception room with great respect. For the first time, it suddenly dawned on me that my card had great impact. On my card my title was listed with the address of the generalissimo's Xian headquarters. The magistrate received me cordially, and I came directly to the point.

The magistrate understood his role. "How many laborers do you need to complete the project in the required time?"

I calculated quickly. To complete the section of road through his county, we would need about one thousand laborers. I added twenty percent for bargaining and another twenty percent for absenteeism, reaching fourteen hundred. Then I rounded the figure upward and asked for fifteen hundred men.

"You are talking about fifteen hundred men and women," he corrected. "We have only twelve hundred families in this small county, excluding those too young and too old to contribute. Each family can spare one man at most for the road work, so the rest will have to be women."

I was surprised, since I had not expected him to consent to my request so readily. However, he was even better than his word, for within two days he had recruited more than two thousand laborers. Our crew finished his section of the road in record time.

I found out later that he greatly admired the generalissimo and would readily give up his life for any cause of Jiang's. I seldom saw such loyalty to the generalissimo.

In the next county I was less fortunate. The magistrate there had broad experience during his term and knew how to evade any proposition that promised him little or no profit. As soon as I mentioned the road work and the necessary local labor commitment, he started an emotional speech that was sad enough to make the marble statue on his desk weep.

"You must understand me. My humble county is very, very poor. You have heard about the bad crop last fall, the plague that killed our livestock, the locusts that ate all our harvest, the flood that inundated our villages, and the bandits who took everything we had left. At present, we have to supply the army with rice, mules, fuel, oats, and whatever else they want. What is more, most of the able-bodied men are joining the army."

I could not get a word in edgewise.

"Now you want labor for the road. Why? The road is good enough as it is. Believe me, I have seen trucks driven through my county, and it took only three days. Yes, I understand this is the generalissimo's order, and I know he wants to fight the Japanese. I have given enough to help his cause. My people cannot suffer anymore." He broke into tears. "Oh, have a heart for my humble county; you can get as many workers as you want from the next county."

I said nothing for a moment.

Finally he whispered, "Do you want to compromise?"

I realized I couldn't reason with him. Instead, I got right to the point. "Look, even though this project is forced labor in principle, the work will be paid for."

He seemed electrified and sat bolt upright, eyes wide open. "How much do I get? How much money have you brought with you?"

"Wait a minute," I said. "Do you agree that in three days you can round up three thousand laborers for the road?"

"Well, that can be arranged if they are well paid," he said.

"I am going to pay them $.10 for each cubic meter of dirt they move, and that will total about $15,000 when the work is completed." I saw in his face that the bait was beginning to work, so I went on. "If you agree to this proposal, you will send the necessary documents to each village for the labor at once!"

His entire demeanor changed. He had the documents ready within the hour. When he was ready to send them, I interrupted.

"I am going to see them sent through personally," I said, taking the documents. His face changed, for he knew that this could have been a fat profit for him. He wanted to send subordinates to each village and round up all the available men. If any of them declined, he would "compromise" by keeping their potential pay. He calculated that he would make a handsome sum from his beloved citizens by "compromising" portions of the amount I handed him.

I decided to go to each village and deal with the people in person. The magistrate tried various ways to discourage me from negotiating with them directly, but I insisted upon doing so. Finally, I threatened to report the entire matter to the generalissimo, and that ended the argument.

The following day I sent my men to each village and assured the people that they would be paid if they completed their share of the work. At first only a few came, for they had never seen any government bureaucrat put good money into their hands. They thought of it as "extracting meat from the jaws of the tiger." However, as soon as the first payment for work was delivered, the news spread. People from all over the county came to work on the road. The project was finished in due time and much more efficiently than I had hoped.

I considered reporting the magistrate to the government office in Chongqing. However, I remembered the old saying "sweep the snow on your own front porch; do not mind the frost on the other's roof." I decided to let the matter drop.

We entered very rough and hilly terrain in the next county. The county administrative center lay in a valley shadowed by a mountain, and when we arrived angry people were crowding the building. I asked about the excitement and was told that the people of the county were rebelling against the county magistrate.

The real cause of the problem lay with the deeply entrenched "bang," or secret society. These societies were organized during the Manzhou Dynasty with the intention of overturning the Manzhou emperors. With the end of the Manzhou, the societies continued to grow. Nobody knew the extent of the membership of different bangs.

When someone became a member, he could get assistance from other members of the bang whenever he was in need. In return, he was obligated to render assistance to other members. The bang organization was based on seniority, and senior members could give instructions to younger members, regardless of whether they were acquainted.

When bang members traveled, they sometimes had to establish contacts in new locations using simple signals. For instance, a member might frequent a public place, such as a teahouse, and place his hat upside down on the table. As a further sign, he then might raise his teacup with both hands, using four fingers of the right hand and three fingers of the left hand. When someone approaches and asks, "What is your honorable name?" he will be holding his teacup the same way. To make sure the signals are intentional and not accidental, more specific signs are required in conversation:

"At home my name is Pan, and outside my name is Li." This might mean the member's name is actually Li and that Pan was the name of the founder of the society.

"Which direction does the main gate of the great city of Changan face?"

"It faces south, with eighteen Buddhas on the gate," would be the required reply.

Sometimes this unusual form of conversation would go on for quite a while. If the visitor could answer all of the questions with no mistakes, the second man would ask the first man's rank in the bang. If the traveler's problem was too great for the second member to help with, he would introduce the first to other bang members. In no instance was a member left without assistance.

The problem we faced in doing road construction was that roughly two-thirds of the local men were bang brothers. For some reason they had united in revolt against the magistrate. When the magistrate received us, all he could say was "look at the mobs outside. This is a fine time for you people to come for laborers. Please do not bother me anymore with your road, as I have trouble enough here."

I could see he was telling the truth. As I left, I knew I would get no laborers from his county. Even worse, this might set a precedent for other counties not to cooperate. My deadline was thirteen days away, and no labor meant no road.

Privately, I consulted my fellow engineers. Baizheng Nei, an experienced railroad man in his forties, volunteered to break the deadlock. Mr. Nei was a man of very few words, but his weather-beaten face and husky build showed he had been many places and knew a great deal about life.

"If it is a problem with the bang, I think I know how to handle it," Nei said. I chose not to ask how; his competence and confidence suggested that he knew what he was doing.

Later I learned that Nei went to a local teahouse, and within half an hour he was ushered in to see the "Dragon Head," the local bang leader. When the Dragon Head found out that Nei's rank was much higher than his own, everything was resolved quickly. All I had to do was leave Nei stationed in the county, and we would get all the labor we wanted.

We passed from one county to another until we had traversed all eleven counties along the length of the road. Using different strategies we coaxed, coerced, or cajoled each county into contributing labor — some with ease, and some with difficulty. At the end of our fourth day, we had twenty-two thousand laborers working on the road.

Organizing twenty-two thousand workers for one project proved to be quite a task, particularly since most of them had no roadwork experience. For example, they needed to have at least five feet of space between them to avoid interfering with one another. As soon as we moved them apart, they would pile together again so that they could not raise their picks without hitting the head of the person behind them. However, their general inefficiency was offset by their long working hours. They worked from the first sign of light and continued until dark. The workers took a break for lunch, which consisted of some wheat cakes and water.

These workers were the sons and daughters of farmers who for generations had been taught to work steadily, but they had never been taught to value efficiency. About the tenth day, I began to see the brighter side of the project. The workers began to get used to the roadwork, and inch by inch the hills were lowered and the road began to take shape. It became less a question of whether we could do the job than one of whether we could finish the road on the specified date.

We tried every trick we could think of to speed up the work. Tobacco and wine were distributed to all who worked especially hard. Those who failed to work with enthusiasm were duly punished.

Finally, I divided the road into short sections and assigned about fifty workers to each of the smaller sections. I posted a dollar, which was a substantial amount in wartime China, on the top of a bamboo post. The group that completed its day's work first got the dollar. The workers were so eager to get the reward that they did a usual day's work in half a day.

Throughout the entire fifteen days, none of us had much sleep. We covered every square inch of the road by foot or on bicycle. To my great relief, the project was completed by the end of the fourteen days. Of course, it was not exactly a first-rate highway but was a graded dirt road, and it could be traveled only in dry weather. Most important, however, we had met our commitment.

The generalissimo's deadline was real. On the fifteenth day, a caravan of about five hundred heavily loaded trucks came tearing down the road, stirring up a cloud of dust the entire day. The laborers who remained near the road cheered them on. They didn't really understand why the road had to be built in such a hurry, or where the trucks were going, but they were glad their job was finished so they could return to their farms.

Actually, the route had historical significance. An ancient roadway had passed along it, beginning in Xian and leading to Sichuan Province. It was an important route during the Tang Dynasty, when Changan was the capital city. Emperor Ming Huang ruled brilliantly during his younger days, but he later became enchanted by his beautiful concubine, Yang Kuefei. This infatuation led to a general neglect of state affairs. General An Lushan appeared to be a very loyal subject, but he secretly plotted against the emperor. When An Lushan marched his troops into Changan and forced the emperor to flee, the emperor and his palace guards escaped along the route of the road we had built.

A short distance from Changan, the imperial guards refused to march any farther. The tired and hungry guards demanded that Yang Kuefei's brother be executed. This was done, but it didn't satisfy the guards because they decided that the root of the entire problem was Yang Kuefei herself. The emperor realized he had no choice, and with evident agony he ordered the death of his beloved concubine. She was strangled before his eyes. The emperor then proceeded to Baoji along the route of our road and entered Sichuan Province, a mountainous area almost inaccessible to invaders.

The emperor is said to have spent the rest of his reign lamenting the death of Yang Kuefei. First narrated by the famous poet Pai Chuyi, this story is known by everyone in China, just as everyone in the West knows the story of Romeo and Juliet. Our road passed directly in front of Yang's grave at Mawei. All of the laborers stood in silence before it, and I ordered a shortened workday in memory of this historic tragedy.

– Chapter Five
Xian to Lanzhou

On July 7, 1937, the Japanese advanced from Manchukuo and attacked China near Beijing. This attack, called the Lugougiao Incident, marked the beginning of eight bitter years of war with Japan. Even worse for China, the generalissimo lost one battle after another, including a bloody battle at Nanjing. It was heartbreaking for me to learn that my family had to evacuate Nanjing with only a few hours notice, leaving everything behind.

We heard rumors that Generalissimo Jiang did not intend to engage his crack forces against the Japanese because he called the Japanese a skin disease, whereas the Communists were a problem of the heart. However, after the fall of Nanjing to the Japanese, the capital had to be moved west to Chongqing in Sichuan Province.

The great efforts of the Chinese Communist forces against the Japanese were little known, but rumors of them reached us at times. We heard that in September 1937, at Bottle-Shaped Pass near the Great Wall, the Eighth Route Army led by Communist General Lin Baoji had ambushed the Itagaki Army, one of the crack Japanese armies. This was a great victory for Lin. Such victories were never reported by the Nationalist government.

This was a time of terrible turmoil, and whereas the Nationalist government had a brief respite in Chongqing, millions of Chinese moved west to escape the marauding Japanese. I considered myself fortunate to be in Xian where everything was relatively peaceful. People went about their daily business as though the war did not exist. My latest assignment was to survey the route of a new highway from Tangchang to the border of Sichuan.

My first task was to drive my crew from Xian to Lanzhou. We made the five-hundred-mile journey to Lanzhou at a rate of about fifty miles a day when the weather was favorable. On the fifth day of our trip, we worked our way over the steep and treacherous slopes of Six Fold Mountain.

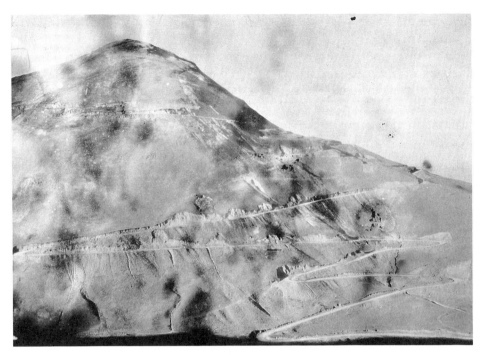

Six Fold Mountain where Kublai Khan died after conquering Europe

According to legend, Six Fold Mountain is where the great Kublai Khan died and was buried. His tomb has never been found. I understand that recently a joint Japanese-Mongolian team of scientists, using remote-sensing technology, found evidence of a humanmade construction underground that may be his tomb.

Past Six Fold Mountain, the mountains were less steep and rugged and were all about the same elevation. They seemed to roll on and on in a continuous wave as far as we could see. The mountains strongly resembled ocean waves. They did not show a piece of rock, a single tree, or even a patch of green turf in the entire range. We saw nothing but the barren yellow loess on all sides. The road went over the crest of the mountain, turning and winding so often that we lost our direction and wondered whether the car was carrying us back to where we had started.

We didn't meet a single person along the way until we came to a small valley in which two humble huts stood. We stopped the car and asked the owner of the huts for some water to fill our overheated radiator. The source of the water was a small, half-dried spring seeping slowly through the dry soil. The rate of seepage was so slow that it took nearly an hour to fill a five-gallon can.

While we waited for the water, we fell into conversation with the owner. He was an old man, and his wife looked even older. "How do you manage to live in this devil land?" I asked.

"By planting corn and potatoes," he answered, pointing to a miserable plantation at the foot of the slope.

"Can your family live on this handful of crops?"

"Well, if God is merciful and sends us enough rain, we have a fair chance. God has not been so generous in recent years. Last year we lived on corn stalks, and this year, if the weather doesn't turn cold too abruptly, we might have some cornmeal."

"Why don't you leave this place and search for better land?" I could see no reason for this old couple to try to hang on in such desolation.

The old man spoke slowly, in bitter recollection. "In my youth, I was ambitious and strong. I managed to get a piece of bottomland near a river. But then a drought came, and then the locusts, and even heavier taxation than before. We were finally driven back here, where my ancestors once lived. Here we have even worse drought and more hunger, but we are free of taxation. Now that I am getting old, I am content to die here and be buried where my ancestors were buried."

We all felt sorry for him. It was, in fact, only one small tragedy in this land of suffering. The old man still possessed his huts and enough clothing to cover his back. I found out later that he was considered to be much better off than thousands of others. We had no way to solve the fundamental problem in the region, which was the lack of water. It could only be brought in through an irrigation system, which required more and better roads. Our roads were a modest beginning, but a great deal of work had to be done.

When the radiator had been refilled, we drove on. In the afternoon the weather changed. A strong north wind was blowing steadily, followed by intermittent rainsqualls. The rain soon turned into snow, and before dark we were in a fierce snowstorm, with steadily dropping temperatures.

That evening the temperature fell to ten degrees below zero. We tried to push on, but the car soon stalled in the deep snow, and we gave up any hope of reaching the next village for shelter. All we could do was unpack our luggage in the car and wrap up in fur coats to wait for dawn.

We had our first real experience with the realities of the great northwestern region of China. The cold night stretched on. As we shuddered in the cold, I could not help thinking of the old man we had encountered only this morning who was so optimistic about his cornmeal.

We reached Lanzhou fifteen days after we left Xian. All of us were very excited, for at last we had reached the heart of the great northwest. We felt we were on a great adventure. To our young men, who had lived all of their lives in the coastal cities, Lanzhou was so remote that they felt more familiar with Chicago or Paris.

Later, when we looked it up on the map, we realized that Lanzhou was situated near the center of China, and there were thousands of miles of land farther west awaiting development. We had traveled so long and yet had covered so little ground. To a group of young engineers, the work seemed endless.

Lanzhou, the capital of Gansu Province, had been a major strategic and commercial center in early times. Since then, much of history had passed it by. Lintao, the western end of the refurbished Great Wall built during the Qing Dynasty, lay a short distance south of Lanzhou.

The Huanghe (Yellow River) runs through Lanzhou. In the winter the local people crossed it on ice sleds. In summer they used sheepskin rafts for transportation. They inflated the sheepskins by blowing into them and tying them at the neck. A number of inflated sheepskins were tied together and overlaid with planks to form a raft. This was the most popular means of river transportation.

In the late 1920s a steel truss bridge was erected across the river. The truss was transported to the site section by section on packmules. Today Lanzhou is a major industrial city. It is the center of the petrochemical industry and the site of China's nuclear facility.

Our journey had not ended. Next we had to go from Lanzhou to Tangchang, which is a small town. This stretch of road was in no condition for motor vehicles, so we had to hire horse-drawn carts to carry the baggage while we traveled on foot.

The march was exhausting, and we covered little distance each day. Of course, there were no hotels along the way, and we were lucky if we could find a small inn where we could get some hot water and a decent meal. Most of the time, we were forced to spend the night in local villages, sleeping on muddy floors.

What troubled us most were the bedbugs that swarmed about us as we slept. When the first batch of bedbugs had sucked enough blood from us, usually from our necks, the next group would take its place. We awoke each morning with swollen necks and red eyes from lack of sleep.

In some villages a few of the engineers managed to obtain cots and hoped to stop the bedbugs by placing the foot of each leg of the cot in a bowl of water on the floor. However, the persistent bedbugs crept up to the ceiling, and with the precision of a bomb sight, would drop onto the cot. In fact, the men on the cots were in worse shape than the rest of us the next morning, since the bedbugs had no way to get off the cot and would feed on them all night long.

We finally reached Tangchang, where the magistrate greeted us. He took us to the best house in town, which was occupied by a wealthy herb dealer. We were given the main chamber for our sleeping quarters. Unfortunately, we had to share the quarters with a coffin containing one of the owner's relatives until burial. Some of the men in our group swore they heard noises in the coffin at night.

– Chapter Six
The Survey Through the Mountains

At last we were in a position to study our proposed route along an existing trail. It was too rough and narrow for modern vehicles here, and we had no idea what kind of country this trail crossed. Asking the local people was of little help, because few of them had traveled more than ten miles from their homes. Those who had traveled greater distances gave only vague descriptions of what they had seen, so we had to reconnoiter the route with little help.

I discovered that we had been assigned to another significant historical route. It had been taken by General Deng Ai when he invaded Chengdu, the capital of Sichuan, in A.D. 200, near the end of the Three Kingdoms period. This difficult trail had never been well established, and no one had suspected that any invader would be foolish enough to lead an entire army along it.

In our time, the trail was used by local people mainly to carry medicinal herbs to Sichuan, and it was even rougher than we had expected. It followed the White Dragon River most of the time, occasionally leading into a canyon with vertical cliffs on both sides of the river. Where the river was forced suddenly into narrows, the current shot furiously through the canyon.

To provide a footpath along the canyon, the locals still used the timber trestle that had been built by General Deng almost two thousand years ago. Obviously, the trestle had been maintained and repaired as necessary. Ancient inscriptions carved into the rock cliffs told the story and gave the date of the construction by General Deng's army.

Theoretically, the road could be built through the canyons by blasting and cutting through the cliffs, but the cost of such construction would be very high. I wanted to know if there was any other possible route through the area. Since none of the local residents could answer that question, I went to the county magistrate.

This county lay in the heart of the mountains and had a population of no more than five hundred people. The magistrate received me cordially and with obvious delight; it turned out that this was the first time since his appointment to the job that anyone from the outside had paid him a visit. As a representative of the national government, I was considered an honored guest.

Before I had time to make my purpose clear, he showered me with a barrage of questions about national events. He did not know that Nanjing had been evacuated by the Guomindang and taken by the Japanese, and he knew nothing of even older news. I brought him up to date the best I could.

"In this county our only source of communication with the outside world was through a radio set," he told me. "Unfortunately, it broke down some time ago, and no one here can repair it."

"What about the road? Is there any other route other than the native herb trail I have been following?" I asked, steering him back to the point.

I learned that the magistrate knew very little about the geography of his county, but he was willing to help. He summoned his office staff, and I tried to learn of any other possible route. No one had any useful information. Finally, I decided that trying the route was better than arguing about it.

The magistrate selected about twenty armed guards from his office staff, a guide, and a sedan chair with four coolies to carry me. It was a royal send-off, with half of the town turning out for our departure. In addition to the company of men, six donkeys carried our luggage, but we did not carry any food; the guide assured us that we would be able to buy meals along the way.

The first day of the trip was splendid, with fine weather and a route through an open valley big enough for a six-lane highway. As we proceeded through the far end of the valley, the trail narrowed. In my mind I reduced the six-lane highway to two lanes, then to a single lane. Finally, the men carrying my sedan had difficulty turning around the sharp bends, so I abandoned my relative comfort and returned to walking.

By the end of the third day, we realized that the country was almost uninhabited. The mountains pressed in close on all sides. Reluctantly, we abandoned the donkeys at a deep gorge where a single log served as a bridge. We had to cross the log by crawling and scooting on our bellies. At this point, most of our guards abandoned us, leaving our guide and a few of the guards. Our party shrank to about a dozen men who crossed the gorge on the log.

I still wanted to observe the entire route, although I saw very little chance that we could put a road through such rough terrain. The next morning it started to drizzle, and by noon it had become a downpour. We had no shelter, and we could do nothing but proceed to the next village, which our guide vaguely remembered was somewhere ahead. As the rain intensified it formed a thick veil of mist over the densely forested mountains and obscured our vision.

The trail became steadily more difficult. Sometimes we would advance for a mile or two, and the guide would decide that we had taken a wrong fork. We had to backtrack, then the guide would decide that the previous route was the right one after all. I said nothing, but I gritted my teeth and walked fast to drive away the chill running down my spine.

We marched all day with empty stomachs and ill tempers. Shortly after 4:00 P.M., the shadows of the surrounding mountains fell across us. Two hours later it was completely dark, and we had to feel our way carefully through the dense undergrowth lest we tumble into some hidden crevice. We finally stopped asking our guide how much longer we had to go, because he clearly had no better idea than we did.

Night fell, and we could hear the hooting of owls. We were cold and damp and were desperate for shelter and did not want to stop to rest. Finally, we heard dogs barking. It was a welcome sound, because where there were dogs there were surely people. We were puzzled, though, because even though we moved closer to the barking dogs, we saw no lights.

Half an hour after we first heard the barking, we walked into a deserted village. Not a single living thing came into view, not even the dogs. We found no fires in the kitchen stoves, no pigs in the sties, no furnishings in the houses, and what was more, not a single grain of rice or ounce of flour. Although we did find signs that the occupants had left hastily and recently, we did not know why.

A deserted village was better than no village. The huts offered shelter, which was far better than sleeping in the rain. We needed to build fires to warm ourselves up and to dry our clothes. The firewood, like everything else outside the huts, had been soaked by the rain, so we tore down some window frames to burn. As the fire glowed and we slowly dried out, we realized that we had eaten nothing since that morning. We had already seen that there was no food in the village, so we searched the nearby rows of crops. Our guide soon unearthed a handful of potatoes, and we baked them by the fire. We had a single-course meal of baked potatoes with no salt.

Now that we were warm and fed, the guide finally relaxed and became eloquent. He began to tell us stories about the people in the region.

"For one thing, salt is a novelty here," he said. "Salt was imported from Sichuan, and it took months to get here. It is a luxury, and only the well-to-do can afford to have it on the table." We thought this might account for the swollen lumps we had seen on the necks of some of the local working people, which resulted from a deficiency of iodine.

"Those who can afford to have some salt use it sparingly," he continued. "Usually, during the meal a small lump of salt is taken out and passed around, and each member of the family licks it. I heard a story about a poor family who happened to possess a small lump of salt. It was so small that it would not last if everyone kept licking it. The father hung it over the table with a piece of string. During each meal, the family members could look at the salt while they ate. One day the eldest son complained, 'Mama, my younger brother has glanced at the salt twice during only a single mouthful of food.'"

The long hike through the cold, wet mountains had exhausted us all. As we became more comfortable, the conversation wound down. Still wondering why the village was empty, we soon fell asleep among the bedbugs and mosquitoes.

We awoke the next morning to find the weather much the same as the day before. It was still raining and very dark. However, some of the elder members of the village cautiously returned.

They explained that the villagers had spotted us long before we reached the village. Since our guards were armed, they had taken us for bandits and had fled to the mountains carrying everything they could. An old man over eighty remarked that this was the first time in his entire life that he had seen an outsider. As we spoke, we put the elders at ease. Some of them signaled for the other villagers to return. They came back carrying their belongings and entered their homes again.

I was wearing a leather jacket, and the villagers were very curious about it. One after another they felt the leather and were surprised at how softly the leather had been tanned. Then they examined every item we had with us, just as a baby examines everything new it sees.

We tried to give them a five-dollar bill for all the trouble we had caused and for the damage we had done. They had no understanding of its value. We tried to explain the equivalent amount of coins to them, but they refused to accept the money, partly on the grounds that if it had the value we claimed it did, it was too much for the window frames we had burned. Finally, we rounded up all the small change we had among us, which totaled considerably less than $5.00, and they accepted it.

The villagers were kind enough to send a guide to conduct us to the next village. Unlike our own guide, he knew his business. This day's trip was much more successful than that of the previous day, although the rain never ceased.

We arrived at the next village before dark. With our guide from the last village to introduce us, we were welcomed. However, we discovered that our dinner here, too, consisted of nothing but boiled potatoes, mashed potatoes, and fried potatoes, still with no salt. The potatoes we had dug up the night before truly represented the local cuisine.

The following days of hiking through the mountains proceeded in much the same way and with the same food. Like the others in our group, I became sick of potatoes; the very sight and even the mention of potatoes made me lose my appetite. We cursed the man who invented the planting of potatoes. We dreamed fondly of even sweet potatoes, since they had flavor even without salt, but it was just a dream.

On the thirteenth day of our trip we plodded into Wudu, the last town along the survey route before we reached the Sichuan border. By that time this small hamlet looked like a real town to us. It consisted of several hundred houses and even had a small general store that stocked commodities such as farm implements, dental fittings, dried beans, flour, brown sugar, black salt, and even cigarettes.

I eagerly seized the cigarettes first. The cigarettes I had been carrying with me since the beginning of the trip had been constantly soaked with rain. The variety of food in Wudu was also a luxury. We refreshed ourselves in the town, relaxed, and spent the night.

When we left, the leader of the town was kind enough to send ten more guards with us. He also sent three sedan chairs, which we really didn't need since we had less than twenty miles to go. Once again, we left town in grand fashion.

At last we reached the spot on the border of Sichuan Province where the rest of our party was waiting. We were comfortable, rested, and filled with great cheer. No one could have guessed what we had experienced to get there.

– *Chapter Seven*
Along the White Dragon River

Now that I knew the herb route was the only possible route for our road, we started the surveying work in earnest. Our day started at 4:00 A.M., long before the autumn sun rose. I was always the first to rise, and by blowing a whistle loudly in everyone's ears, I forced the others out of bed.

Our cook was a very sturdy man with short, husky limbs. His appearance had earned him the nickname "Gorilla." Gorilla was the heaviest sleeper I had ever seen. Since he was the cook, he was supposed to be the first to rise in the morning, but this never occurred.

Blowing the whistle, shaking him, and speaking to him did not wake him. When I had exhausted every other means, I poured cold water over his head. It was the only thing that worked. I once found him snoring by the stove with a hot fire going that was so close to him that his eyelashes were singed, but he still didn't wake up.

Our breakfast that first morning consisted of rice soup, wheat cakes, and plenty of eggs. After breakfast I blew the whistle again, and we started for the road, even though it was still dark outside. As we walked, predawn light lit our way.

Beautiful landscape surrounded us. We trod through the fields of corn and other crops saturated with morning dew, and soon our pants were wet from the knees down. Still, the morning was so pleasant that we forgot our recent misery. Even better, we had no newspapers or radio announcements to bring us news of the war.

As we crossed the tilled fields, the peasants coming out to work eyed us suspiciously, wondering why townspeople had to get up so early. They did not bother us, however. When we reached the surveying route about a mile from camp, the sun was just peeping over the horizon.

We divided into small surveying parties, and our work progressed smoothly, with members competing with each other to be the most efficient.

The instrument man lost no time in sighting his target, and the chain man carried the chain along his route at a run. The stake man swung his ax high enough to drive the stake flush with the ground in one stroke.

Each day proceeded about the same as the previous ones. All the strenuous exercise gave us huge appetites. Our large breakfasts were burned up in no time, and we would look for the cook, who was scheduled to send us lunch at midday. Lunch was always late, however, because the cook took a morning nap after he had finished preparing breakfast. One day, after waiting our entire lunch hour without any sign of the cook, we picked some corn from a field, built a fire, and roasted the corn to a yellowish brown. It was delicious.

After we had eaten, our tardy cook showed up staggering. I swore he was drunk and hurried to him, but I discovered his most remarkable stunt yet. He was virtually asleep while he was walking, and his eyes were closed.

The surveying work continued until the sun set in the west and darkness began to envelop us. Then we would drag our weary feet back to camp for supper. Since we left the camp before dawn and arrived back after dark, we never saw the camp in daylight. The long work schedule sometimes reached twelve hours a day, but nobody complained since we were all eager to get the job done.

We soon passed through the relatively limited area of level ground within our survey route and reached the rough country. Here we followed the course of the White Dragon River very closely. Our line would now have to cut through the ledge of a canyon and the side of a hill. All of our talents and physical endurance were tested to the utmost. Nearly every day we had to scale steep cliffs and wade through swift mountain streams that fed the river.

At this stage we quickly discovered that our leather shoes were unsuitable for climbing here. We changed into the footwear local peasants wore, sandals woven from straw. These sandals saved us from many false steps because they would grip even on a slippery rock.

Wading across the river and streams was not nearly as easy as it looked. Where the water was neck high, it was safe to swim across. If it was only knee-high, swimming was too dangerous because of rough boulders underwater. Looking down at the water for several seconds while we were wading made us dizzy and disoriented. The best way to wade was to keep our heads up and look at the other bank while taking slow, careful steps so we could feel the riverbed with our feet.

In the flat country we had covered about two miles a day, but in the canyons we were lucky to travel half a mile a day. Thus for months we struggled through the mountains and across rivers and streams. Our senses of hearing became dulled by the constant thunderous roaring of the river. Even our vision of the sky altered from a panorama of open space that stretched endlessly overhead to a narrow strip directly over us that was bounded by towering mountains on each side. The sun rose later and set earlier because of the mountains.

The rougher the country, the poorer the inhabitants became. Farmers rarely had more than several acres of level ground. Still, the people managed to get by, building their houses on the edges of rock cliffs to avoid wasting any of the precious arable land. The houses were usually two stories high with flat roofs. The first floor sheltered livestock such as pigs, sheep, and mules, and the second floor housed the family, although it reeked with the smell of animal manure. Even the flat roof held shallow soil for growing a crop.

It was well-known that the Chinese respect the dead. Ancestors were generally buried as lavishly as possible, and great care was usually taken by both rich and poor to leave the family graves unmolested. Here, people did not even have enough land for crops, so they could not sacrifice any of the land for graves. They simply stuck the rough coffin in a crevice in the rocks, where it was out of the reach of molesters, and left it there. It was not unusual during the course of the survey for us to find a coffin that had rotted, leaving the legs of the corpse dangling over our heads.

As poor and limited as the land was, the peasants had an excellent yield, especially of fruits. Walnut, pear, peach, and persimmon trees thrived near the river and ripened while we surveyed the route. Since the peasants had no means of transporting their crops to a market, the fruit cost almost nothing. For a dollar we could have as many pears as we could carry. This constituted one of our chief pleasures, and no canned fruit could match the flavor.

One day we arrived at a lonely village located at a bend of the White Dragon River, where both sides of the river were hemmed in by high, formidable cliffs. The rocks were a deep-red color, and as the evening sun reflected off them the even deeper red was almost blinding. No painter could have duplicated the color; nor does any description do it justice. Color film was not readily available at that time, so we had no way to preserve the sight.

During our short stay in the village, we all developed an uneasy feeling of foreboding about the towering cliffs, as though the surrounding mountains would rise up and smash us to pieces. When we mentioned to the villagers that the cliff appeared ready to fall, they informed us that the mountain on the other side of the White Dragon River was an extinct volcano. Several years later I heard that a tremendous landslide there completely dammed the narrow riverbed and flooded the banks for miles upstream, killing hundreds of people.

One night I was on the flat roof of the house where we were staying. I could see a red glow about fifty feet high in the thick forest at some distance. At first I thought it had to be a forest fire, but the yellowish-red color would flare up for half a minute, disappear completely, then flare up again; a forest fire would not shed light that way.

Since that part of the mountain was insurmountable, I was sure no people could be up there. I woke the crew and the peasants in the village to witness the strange sight. The locals were clearly awed, which meant they had no simple explanation either. I could understand their reaction because an ancient Chinese folk tale tells of something like this phenomenon.

According to the folk tale, the fox was so clever that it could acquire human knowledge. In fact, magical powers allowed foxes to change into human form. However, to accomplish the last stage of the transformation, the fairy fox had to extract the essence and beauty of the moon and work it into a small pellet, a process that could take several hundred years. If the pellet was stolen by a human, all the work of the fox fairy would be lost, but the human who swallowed the pellet would have eternal life. For this reason, fox fairies jealously guarded their treasure from all intruders and chose secluded places to finish making the pellets. The red glow, as explained by the local peasants, was the work of a fox fairy who was on the verge of completing a magical pellet.

Of course, that explanation did not satisfy a crew of hardheaded engineers, but we failed to come up with another explanation. Unwilling to dismiss the question so easily, some of us set out to explore the area of the red glow, despite the strong objections of the natives. We gathered all the flashlights we could find and some torches and left immediately. After several hours of rough climbing, our search party reached the foot of the mountain from which the red light had glowed. By that time, however, dawn had broken, which washed the red light out completely. We had accomplished nothing except losing a night's sleep. Even now, I cannot give a likely explanation for the entire episode. The light may have been caused by volcanic activity, although the volcano in the region was supposedly extinct; it also may have been a UFO. The light did not appear again during our few remaining days in the village. The incident did serve to remind us engineers that there are phenomena in this universe that cannot be readily explained by science. Seeing is believing, and one cannot always discredit tales heard or events read until one actually witnesses the particular phenomenon.

There was no doctor within a hundred miles of our location. Fortunately, no one in the crew became seriously ill. On one occasion I developed diarrhea, and the only person in the nearest village who could help was an acupuncturist. I had no choice but to visit him. He assured me that the problem was not serious and produced a long, thin, rusty needle. Even worse, he sharpened the needle on his mud-covered shoe. I doubted he had ever heard of sterilization. However, before I could react, he stuck the needle in my stomach, right through my clothes. He gave it a few quick twists and pulled it out. Miraculously, my trouble vanished, but I would never try that treatment again.

Three months after we started the survey, we had completed most of the assigned route when we reached the foot of Touching Heaven Mountain. In A.D. 200, when General Deng Ai followed our survey route to invade Sichuan Province, he was halted by this mountain. Deng's army could only take the city of Chengdu by surprise if it could cross the mountain. When Deng's troops refused to proceed down a steep slope, Deng Ai ordered his soldiers to wrap themselves in heavy blankets and roll down the steep cliff. Deng went first, and his men followed as ordered. They captured Chengdu and became known as the "Army From Heaven."

In our time a thick forest covered the entire mountain. The trees were mainly bamboo, and the growth was so dense that it blocked the sun. Even in daytime we sometimes had to use torches to find our way. The herb trail was so steep and narrow that it was not usable for highway construction.

I divided my men into several parties, each assigned to scout a possible route. By the end of the week I realized I had accomplished nothing. We could not see even fifty feet ahead in the thick growth.

We set up our camp at the top of the mountain, around eight thousand feet above sea level, so we could go in any direction. The site was cold and wet, and our food supplies were getting low. Every day, the crew went out through rain, frequently without food. They were becoming so miserable and dejected that I finally decided it was useless to go on this way. The time had come for me to report the situation to the government in Chongqing. I wrote a telegram describing our progress and our difficulties, stating the conclusion that the route was impassable. The telegram was sent by a man on foot to the nearest post office, which was three days away. We expected to hear from Chongqing within a week and in the meantime took a much-needed rest, our first in three months.

Ten days passed, and the long-awaited telegram finally arrived. We were excited, anticipating orders to leave that miserable camp and return to civilization. As the telegram was decoded, the faces of the crew members fell.

"No route is impassable, continue the survey, will send more money." The message was brief and authoritative. As Chief engineer Liu had told me, the generalissimo's orders left us no choice: we would have to do what appeared impossible. That night I held a conference in the camp to decide what we would do next.

"What the hell do we care about the Chongqing decision?" one man remarked. "Those people who sit in the Chongqing office have no idea what kind of life we are living here."

"Ask those people in Chongqing to come and see what they can do about it," said another.

"I don't care whether we continue the survey or not," said a third. "I just want to see what we're doing instead of groping around in the dark."

Despair and impatience were making the crew angry, and it was hard to reason with them. However, Nei, a more experienced man, tried to soften the situation. He started by blaming the inconsiderateness of the Chongqing bureaucrats and expressed sympathy for the men. Finally he said, "An order is an order. Don't forget that this is wartime, and none of you wants to be called a deserter."

I finally got everyone to turn their attention to work again. As long as the forest was in our way, we could not select a route. We had two choices: start a fire and burn down the forest, or cut our way through by hand. Since that month was so wet we even had trouble lighting the stove, we would have to cut down the trees. However, we could not do it alone.

When we received the extra money from Chongqing, we hired roughly a hundred laborers from nearby villages. They were assigned to clear a proposed route. As they worked we found the situation was not as bad as we had thought. After two weeks of clearing, we began to establish an acceptable route. After that, work became easier, but we still had a tough time getting around in the survey area. The cut bamboo left a bladelike stump that was sharp enough to cut through our clothing. By the end of the surveying none of us had escaped without some scars.

We struggled for two months in the bitter cold and miserable dampness on "Touching-Heaven Mountain." When we reached the last stretch of the route, we became eager for the work to end. The sun set when we had only a short distance left to go, and the crew didn't want to spend even one more night on the mountain. Rather than return the next day they continued to survey, using a lantern for a target. Close to midnight they finished the work.

At last, with light hearts we descended the mountain. We spent a day in a nearby village relaxing and drying out. When we had rested, we continued on the trail back to Lanzhou.

– *Chapter Eight*
The Gansu-Sichuan Highway

When we had first arrived in Lanzhou, we thought the city was so poor that we would only want to stay a couple of days. Now that we had experienced the bitter side of the northwest, Lanzhou represented paradise. We tried to cover the five hundred miles back to Lanzhou as fast as we could.

"You know what I am going to do first when I get to Lanzhou?" one man asked. "I am going to sleep all day, and nobody will blow a whistle in my ear."

Nei asked, "Don't you want to take a hot bath and wash away the filth that has accumulated on your body over the past six months?"

Another spoke up. "I am fed up with all those undernourished women in the river valley. I want to see some pretty girls, with pretty painted faces."

We all laughed heartily with him. As for me, I wanted to do all those things, too. I especially wanted to buy a pack of Camels and have a real smoke, instead of smoking local cigarettes that tasted like straw.

We reached a small town only a hundred miles from Lanzhou and found a telegram waiting for us from Chongqing. My spirits fell as I read it: "Your party has no reason to return to Lanzhou without orders. Return to the field, and start the construction of the road immediately." The generalissimo wanted the road all the way from Lanzhou to Sichuan to be open for traffic within six months.

I cursed, and everyone else cursed, but we had to follow our orders. In this mountainous terrain, Lanzhou was already within sight, and it was not easy to turn away. I suggested that half of the crew should go on to Lanzhou for a short leave, and the other half would immediately go back to work. Everyone refused, saying that we should continue to share our hardships. So the next morning, our dejected little band turned slowly to retrace the route we had taken only the previous day.

On the eve of the Chinese New Year we returned to Tangchang, which was in the middle of the proposed route. That evening, all the roadside inns had

48

closed for the annual festival. No other travelers were on the road, because they had planned their holiday. Villagers peeked out of their doors at the crazy engineering crew walking down the street when everyone else, rich and poor, was enjoying a warm meal at home.

We faced nearly a month of preparation. We had to organize the construction force, recruit laborers from every county along the line, and transport the construction materials. By the end of February, we had completed these preliminaries.

Tangchang was surrounded on three sides by steep mountains and was bounded on the fourth side by the swift White Dragon River. An old timber suspension bridge spanned the river and provided the only access to the town from the proposed route. The scenery was beautiful, and after almost a year of nomadic life, we enjoyed the temporary stability of living in the town. For a change, we did not have to roll up our bedding every morning and unpack every evening.

With all of the rock work facing us, we could not expect to get the job done with only the recruited labor from the local counties as we had done on previous jobs. This time we would have the local labor do all of the earth work and rely on professional stonecutters to do the rest. The contract for the rock work was put up for bid.

Even the lowest bidder had to bring men from as far away as Xian. Pack teams were used to transport the drills and dynamite. All of the preparation went slowly, and for nearly two months we had nothing to do in Tangchang but sit and wait. Finally, by the end of April, the first epic of stone was cut. From then on the work went smoothly and on schedule.

We had about two thousand recruited laborers and five hundred contracted workers. The engineers spent most of our time organizing all the workers. The local recruits were scheduled to be on the job for at least one year, so we had to gain their confidence. It became part of our job to settle disputes among them. After we had settled several cases fairly, they began to come to us, rather than going to the local magistrate, with their disputes.

We could not compromise with anyone about the route our road would take. Since the country was very hilly, with only small stretches of level ground, we had to take advantage of all the level stretches we could. Landowners naturally did not like this, even though we gave them some compensation.

At first they tried to bribe us with chickens and eggs. When that didn't work, they began removing our survey stakes at night. It soon became commonplace for our crews to drive the stakes during the day and find them gone the next morning.

The laborers received their daily payment in brand-new dollar bills, and the farmers were also compensated without delay. We got the money from the central bank in Lanzhou, and because the bills were crisp and fresh from the mint, the natives were reluctant to part with them.

At one point rumors started that we had our own printing machines and could turn out as many bills as we needed. This was not good, because people began to look upon us with envy. They thought that since we had such a vast supply of money, we were rich enough to help ourselves to whatever we wanted.

Actually, our government pay was lean. I was drawing $220 a month, the equivalent of about 70 U.S. dollars. However, I did not complain, because I was looking ahead to becoming the chief engineer someday with a lofty salary of $600 a month.

One night just before daybreak, I heard the sharp report of gunshots. I rose quickly and, with the others, climbed to the top of the high wall around the house to see what was happening. In the darkness we could make out very little, but from the noise we judged that at least a dozen armed men on horseback were galloping in our direction. We quickly emptied the safe. As the hoofbeats drew closer, we took all the cash and important documents to a corner of the backyard and covered everything with litter. We had barely finished when the bandits reached our gate. Shooting and yelling, they tried to force the door, but it was fairly strong, and they were unable to break in.

Again, Nei took control of the situation. "Don't worry," he said calmly. "Listen carefully. They don't have automatic weapons; they only have shotguns." He distributed three automatics and two revolvers we kept on hand and stationed those who were armed on the top of the wall at each corner.

Other than Nei and me, none of the crew had experience with firearms. Even so, we all fired blindly into the darkness. At the sound of the automatics, the bandits dispersed, only to redouble their efforts at the gate. One tried to scale the wall near me, and I brought him down with one shot. I owed my marksmanship skill to the ROTC training I had received at the University of Michigan.

We remained in position all night. At the first sign of light, the bandits realized we could see them clearly now. They turned away and plundered the town as they retreated.

Thanks to Nei's quick decisions we suffered no losses, but I learned a lesson. We never again displayed the money so carelessly. As a further precaution, I asked for armed guards, but I knew that none of the guards were better marksmen than Nei and that they would be unable to employ any better

Cutting a road through a vertical cliff

tactics than he had at the critical moment. The bandits did not return, and we never learned who they were or from where they had come.

After months of steady work cutting rock and clearing away debris, the construction party reached the most difficult part of the work. This was a location where General Deng Ai had been unable to clear a trail in the usual manner and had built trestles by inserting timbers into the solid rock wall.

We had to blast a groove along the side of the cliff. This would be a half-tunnel, wide enough for one lane of traffic. Our crew had no modern tools, only some steel chisels for drilling and some locally made yellow dynamite.

The chief stonecutter from Xian had no experience in such work, and he could not come up with a method even to get started. The vertical cliff face offered no footing for the workers, and it dropped directly into the roaring torrent of the White Dragon River. With one misstep a worker would lose his life in the mountain current.

We finally came up with a primitive device similar to that used by the Chinese immigrant workers who had built the Central Pacific Railroad through the Sierra Nevada. They had lowered workers in baskets by ropes from the tops of cliffs. We suspended boards similar to a modern window cleaner's scaffold.

At first the laborers refused to be lowered on these platforms, but when we promised them triple pay, they agreed to try. The system worked. The men stood on the swinging boards while drilling into the rock. The three-quarter-inch steel bits sank inch by inch into the rock, driven by eight-pound hammers swung with both hands while the men balanced themselves on the boards. This feat seemed impossible at first, but soon it became routine. Within two weeks the contractor declared that we had enough holes to load the first charge of dynamite.

Initially, the local villagers watched us with interest and sneered at our apparent foolishness for trying to cut a road through the gigantic cliff. They thought we would eventually give up. However, when they saw us loading the dynamite, they were terrified. They came to me as the engineer in charge.

"Don't try to blast the cliff," one shouted in panic. "Those are sacred cliffs, untouchable, and they protect hundreds of travelers passing through the gorge."

I answered calmly, "As far as I know, the cliffs are endangering travelers rather than protecting them. We want the road opened, and anything that stands in our way must be removed."

An elder member of the group talked to me sternly. "Please follow me, and I will show you the will of the god of the cliff."

I followed just to humor him, and he led me through a winding, tortuous flight of stone steps not very far from the cliff. After ascending perhaps two hundred steep steps, we came to a small level area. A small shrine had been cut into the rock.

"This temple was built at the time of General Deng," the old man said. "After his successful campaign, he built this temple to show his gratitude to the goddess of the cliff."

I looked around. The temple stood at the middle of the cliff and had a wonderful view of the gorge. By looking over the edge of the cliff from the temple, I could see our workers below me, industriously cutting the rock.

The old man entered the temple and solemnly performed four kowtows before the stone image, which appeared to be Guanyin, the Goddess of Mercy. Then he handed me a bamboo cylinder that contained several dozen bamboo sticks.

"Here, young man," he said. "Draw one of the bamboo sticks from the cylinder, and see what the goddess has to say to you."

I drew one of the sticks to satisfy the old man and noticed that each stick had a number inscribed on it. I had number fourteen.

The old man took a book from the altar. Fumbling through the book, he located the number fourteen and gave me the interpretation of the lot I had chosen.

To my surprise, the interpretation started with a heavy line marking "disaster, disaster, disaster." The old man read a prediction that anything I tried to do here would lead to disaster.

"What did I tell you?" the old man spoke triumphantly. "So it is for your own good instead of ours that you should discontinue this profitless enterprise."

I was startled by this revelation, but I could not bring myself to believe in omens. "This is all very well, since it's my fate that is at risk. I will stick to my course of action."

The old man sighed and left the temple. He went to tell his people to stay away from us and let us reap our own fate. I went back to work.

When the first charge of the dynamite was finally fired, the report echoed up and down the canyon for miles. Great chunks of rock fell into the river, and dust and foam roiled up from the current. This first blast confirmed our procedure, because it moved the rock as we had wanted. We estimated that with one hundred more such blasts, the roadway would be cut. This would take more than three months.

Maybe I was too optimistic or perhaps just too inexperienced. I failed to notice a fault at the face of the cliff that extended vertically well beyond the temple. One day I was congratulating myself on how well the work was going, estimating that we were about half finished. I had just spoken with the foreman about three hundred yards from the cliff. A thunderous crash echoed around us, as though all of the bombs on a B-29 had gone off at once.

Even before I looked I knew the worst had happened — an engineer's worst nightmare. The entire face of the cliff had come down, blocking up half of the gorge. None of the two dozen workers on the job at the time survived.

This was a great blow. I wired the Highway Administration and told them of the incident, explaining that I could have avoided the disaster if I had been more experienced. The Highway Administration did not take the affair as seriously as I did. I was instructed to continue the work despite the difficulties.

At the same time, the local villagers spoke triumphantly of the disaster, which had been predicted at the temple. I quickly became tired of hearing their constant "What did we tell them?" In any case, the work would not stop.

Finally, I came up with a response to the villagers. "Well, since the goddess of the cliff could not save herself and is now in the canyon with the rest of her cliff, there is nothing to bother us anymore."

Within a couple of weeks we had resumed work. The fallen rock had narrowed the gorge and now formed a gentle slope on the cliff. As a result, our work was easier. I was able to shorten the half-tunnel and clear the rest of the way with an open cut.

In one month we conquered the canyon, although at great cost. We overcame an obstacle that had seemed unsolvable at the outset. I was beginning to feel like a real engineer.

After eight months of hard work, the road opened to traffic. We had built the road within the time limit set by the generalissimo. Of course, we had only built a dirt road, but it would serve the traffic needs of the Chongqing government at the time.

Proudly, I wired the news of our success to Chongqing, and the reply came quickly. Our work was apparently taken seriously, for they sent a high-level government official and a senior engineer from Lanzhou to inspect the road. We made the necessary preparations for the inspection.

I was to drive out and meet the inspection party. To our dismay, on the day of their arrival, a heavy rain fell. The dirt road became soaked, and if we drove the heavy truck on the wet and loose roadbed, we would be asking for trouble.

The inspector's party arrived in the evening. The inspector himself did not make a strong first impression. He seemed to ignore all of the work we had done in the past several months and assumed a stately air, looking down on us as though we were his servants. Since he was the one who would make the report to the government, I dared not show my resentment.

I told him the road was only usable by traffic in dry weather, and with the pouring rain I doubted we could make the trip in the heavy truck.

"Why, highways are supposed to be open to traffic in all weather," he sneered. "When I was in the United States I used to travel in my sedan at sixty miles per hour. We should be able to cover the 150 miles of your road in three hours at most."

He clearly had no experience with roadwork in China and was only familiar with the concrete boulevards in the United States. The next morning the rain had subsided to a drizzle. I tried again to persuade him to postpone the trip until the dirt road had dried out.

"I have important business in Lanzhou, and I have no time to waste on this inspection," he replied. We had to begin the journey.

The first stretch of the trip covered easy ground, but the road remained muddy from the past forty-eight hours of rain. It was slow going for the truck. We covered about ten miles an hour, and the driver often had to use low gear.

The inspector was thoroughly bored before we had covered the first fifty miles. He wanted to turn back to Lanzhou. "I cannot waste my time here any longer. This is the most disappointing trip I have ever taken."

His remarks hurt me deeply. I swallowed hard and thought of the crew waiting cheerfully at Tangchang for the arrival of the truck and some words of congratulation. Finally, the inspector's engineer spoke up in my defense. He told the bureaucrat that he could not expect to travel fast on a new road in this kind of weather. That ended the conversation.

By noon we had gone sixty miles over the muddy road and arrived at a small village where I had arranged for a substantial meal. By now I also thought if the inspector had his fat belly filled, he might calm down a bit. To the contrary, after he had finished his meal, he ordered the truck to turn around and head back to Lanzhou. I lost all my patience and reserve and demanded to know the reason for turning around.

"There is nothing further to see but dirt and mud, and this road is certainly not complete," he answered. "My time is precious, and I cannot delay any longer."

"Yes, all there is to see is dirt and mud, but keep in mind that this dirt and mud represent the blood and tears of thousands of laborers." I was furious and no longer cared whether I offended him or what he reported to Chongqing.

The sympathetic engineer again tried to persuade him to finish the trip, but the inspector's mind was made up. He got on the truck without a backward glance.

For nearly two years we had worked hard in this godforsaken country, and this was our reward. After all the effort we had put into building the road, this soft-bellied man from the city could not be bothered to spend even a day looking at our work. My sense of justice made me so mad I wanted to shout. In fact, I was so angry I did not notice at first that the inspector's engineer had not left in the truck with the inspector but had stayed with us.

"Pay no attention to that thick-skulled good-for-nothing," he said. "I know him very well, and the sum of his knowledge is that he has been to the United States. He gained his position mainly by flattering his superiors and criticizing his subordinates. I'll tell you what I'm going to do: I will inspect the road with you on foot, and I'll file a separate report to Chongqing."

We made the rest of the journey together, and he spoke highly of my work from an engineering standpoint. His no-nonsense approach saved me from a potentially bad situation. I found out later that the inspector had shown no mercy in his report, and had it not been for the engineer's contradictory report, Chongqing might have thought we were nothing but liars and cheats.

For weeks, as I escorted the government engineer along our highway, I could not forget the inspector's irresponsible behavior. My disappointment in him would not abate. Finally, Nei tried to bring me out of my depression. "My brother, "he said, "you are still ignorant of the complexities of the Chinese social structure. You plant corn, but you do not always harvest corn. In this life you must work just for the sake of working. I have long ago given up the idea of fair compensation."

Even though Nei may have been right, it was still bitter medicine. This experience did not match the idealistic notions with which I had begun highway work with the intent of helping to improve my country. I now realized that in China engineering challenges were not confined to engineering itself.

I had always been physically fit and had never missed a day of work because of illness. Now that the work was done, I suddenly felt very tired. At first I thought I was just depressed.

Even though we had completed the highway, we were not to leave until we received government approval of our work. During this time I hated to get up in the morning, and I could barely muster enough energy to walk around my room. At first I thought I would be better in a couple of days. Then one morning I looked in the mirror and found that my eyes had taken on a brownish cast, and my face had a yellow tint. To the best of my medical knowledge, these signs indicated jaundice.

We still had no doctors. With no professional medical advice or treatment available, I decided to head for Lanzhou, which was roughly five hundred miles away, before I became too weak to make the trip. I said good-bye to the crew and left Nei in charge.

For the journey, I hired a sedan chair with four carriers. During the trip I was so weak that I took little nourishment other than water. Fortunately, I had been in good physical condition before getting sick, and I reached Lanzhou without mishap. I had thought I would find proper medical treatment in Lanzhou, but my friends were unable to locate a doctor. All they could find was a trained male nurse to take care of me while the jaundice ran its natural course. I stayed in bed for another ten days. By this time I had seen enough of the great northwest. I needed drastic changes, both physically and mentally. When my condition seemed somewhat improved, I decided to fly to Chongqing on a cargo plane for a complete checkup.

– Chapter Nine
Chongqing and the Southwest

Chongqing became China's wartime capital after the Japanese captured Nanjing. High in the mountains of Sichuan Province, the city was built on the point of land where the Yangzie and Jialing Rivers flowed together. Its position protected it from attack by land. For centuries poets had written about the city. Now it was being touted by statesmen and foreign correspondents all over the world as the Chinese stronghold against Japan.

I arrived in Chongqing in winter 1940. I had recently enjoyed the bright sunshine and the cool, fresh air of the northwest, so Chongqing turned out to be a miserable place for me. Constant mist and fog enveloped the city so thickly that one could barely breathe. At times, the fog was so thick that pedestrians had to carry torches on the streets in the daytime. However, I had passed the worst period of my jaundice, and my checkup showed I only needed more rest. In the city, free from physical stress, I recovered quickly.

During the winter months the sun appeared to be a faint ball of cheese in the sky that only showed itself at midday. The city reminded me of London in that respect, but I hadn't minded the fog in London as much because the mild summer helped to make up for the miserable, foggy winters. Here in Chongqing, although the winter months were constantly damp, it didn't seem cold enough for a warm fire.

Just as I had almost adjusted to the cold damp of winter, summer arrived without much of a spring season. Each day the thermometer quickly climbed past ninety degrees Fahrenheit. The temperature usually stayed above a hundred degrees for most of the day. In this humid climate I became accustomed to sweating day and night and was never able to dry off.

The local residents didn't seem to mind the weather. In fact, nature had provided them with ideal land for crops. The fields produced two crops a year and enough rice to feed everyone in the province and leave a surplus to sell to neighboring provinces. Consequently, the people of Sichuan did not have to work nearly as hard as the people in the northwest, for whom subsistence involved a daily struggle with nature.

I also found that all of the men in Chongqing, young and old, had pipes hanging from their belts. After a couple of hours of work in the fields, they would rest under a tree and smoke their pipes. They claimed tobacco drove away the dampness and even prevented malaria.

Since the capital had been moved from Nanjing, the population of Chongqing had grown by a factor of three. Virtually overnight, thousands of houses had been erected. Primitive bamboo huts with timber and lath walls were jammed against the stately mansions of former warlords who had also been driven here by the Japanese. With such severe crowding, an entire family of four or five people would often live in a single room.

The Japanese had not been able to invade Chongqing by land, but they wasted no time bombing the city. The limited Chinese air force could offer little resistance. However, Chongqing was built on hillsides, and over a thousand air-raid shelters were quickly blasted from the underlying sandstone formation. Chinese intelligence in the outlying countryside was able to inform Chongqing authorities of the departure of Japanese planes from Hangzhou on their way to attack the defenseless city. Officials hung up red lanterns to warn the public, sending us to our assigned dugouts. These exercises became a daily routine.

On June 5, 1941, a tragedy occurred. Several thousand people had taken refuge in the largest bomb shelter, which was about a mile and a half long. A misunderstanding occurred within the air-raid signal system. Inside the tunnel, people thought the "all clear" signal had been given and started to leave. Simultaneously, another signal was given outside announcing another Japanese air raid. Those leaving the shelter collided with those hurrying in, triggering a massive panic. In the confusion, people were trampled and suffocated. Approximately four thousand people died during the incident.

The people of Chongqing were deeply shocked and angered and demanded to know who was responsible for this tragedy. Their grief also generated wild stories. For instance, one evening a housewife supposedly heard a peddler who was selling snacks ringing his bell on the street. She went outside to order a snack, but the peddler had no head. Similar stories were told by many people in Chongqing for months after the tragedy.

The Japanese changed tactics that summer, switching from explosive bombs to incendiaries. All of the makeshift bamboo huts caught fire immediately. In the narrow pathways and with limited fire-fighting equipment, Chongqing ignited one night into a sea of fire. More than two-thirds of the city turned to ashes.

Although the people of Chongqing suffered greatly, their general morale remained high. The brutality of the enemy could not dampen the spirit of the Chinese. This was crucial, since China's war arsenal compared miserably with that of Japan.

In 1992, fifty years after the fact, the Japanese emperor visited Beijing and was received courteously. The emperor said he considered the entire war an unfortunate incident. Nothing was said by either side about war debts or compensation. I witnessed many of the events of the war and have pondered the brutalities of the Japanese, who murdered millions of Chinese. Interestingly, I have rarely heard any protest from the Chinese people, either on the mainland or in Taiwan. However, the Chinese have a long cultural memory and sense of history, and I firmly believe that someday the account will be settled. I do worry that the younger generation of Chinese seems to have very little knowledge of the events of fifty years ago.

When I had recovered fully from the jaundice, I visited the Highway Administration at the capital. The offices, once housed in a grand building occupying an entire block in Nanjing, were now reduced to a dozen simple bamboo huts that together housed over a hundred bureaucrats. Throughout, three employees had to share one desk. This organization controlled all of the highways in China and continued to give orders to men working in the northwest.

I called first on Zhao Zukang, who was the commissioner of the Highway Administration. He allowed me only a brief visit to report on the job I had worked on for the past two years, and he had virtually nothing to say when I finished. However, he did ask me what kind of work I was planning next.

I told him I had had enough of fieldwork and was eager to learn about the functioning of the central administration of China's system of highways. The commissioner appointed me to work in the engineering division.

Naturally, I expected to do some designing and planning in this division. Much to my surprise, little of such work was done there. Rather, the engineering section was kept fully occupied handling the documents of various offices, including the Ministry of War, the Ministry of Resources, and the provincial governments, as well as wires and reports from hundreds of engineering administrators all over the country. Several hundred documents arrived in the office each day.

The documents were remarkably similar. The ministry of war wanted more and better roads as soon as possible to move military traffic; the engineering administrations, which did the actual construction, always wanted more money. Since the government never had enough money to go around, the funds were allocated to the most urgent project. The less important roads would have to wait.

After working in the office for three months, I had begun to learn the essential tricks of the Chongqing merry-go-round. The motto of the bureaucrats was "shift the responsibility to others as much as possible!" We had four standard methods of handling the various documents: present the matter to our superior without comment, pass any orders to the appropriate engineering administration without regard for whether it could accomplish the order, send copies of the document to other administrators and ask for comments, or stash minor documents in a dead file and ignore them.

Needless to say, when we asked for comments from other administrators it took some time to hear from everyone. That was the point. With all of the delays, people on the site often resolved the problem themselves.

When the document was too important to fall into any of the four categories, we could still shift the responsibility by passing it along to other sections of the administration. With each additional chop (Chinese signature), one's own responsibility was greatly reduced. Frequently, when a document had made the long journey through the bureaucracy, it was completely covered with chops. The result was a shared responsibility if anything went wrong.

With this ludicrously inefficient form of administration, I often wondered how any of the engineers at the various posts could complete any construction on schedule. I later found out that the engineers on the job were used to solving problems without any assistance and reported to the administration only as a formality.

By this time Washington had begun to realize that the Sino-Japanese War would not remain a local conflict and was threatening to become part of a growing global war. Japan already controlled most of China's coastal cities, blocking military supplies from the outside from reaching Chongqing.

As a result, military supplies only reached the Chinese troops in western China on the already overtaxed narrow-gauge railroad that ran to Yunnan Province in southwest China from French Indochina — the area that later became modern Vietnam, Laos, and Cambodia. The attitude of France toward Japan regarding French Indochina was unclear; therefore, this single link between China and the outside world could close at any time.

To alleviate this problem, two major projects were under consideration. The most urgent of these was the construction of the Burma Road, a highway that would extend from Yunnan Province across north Burma to India. The second project was the construction of a highway along the Red River, paralleling the narrow-gauge Indochina-Yunnan Railroad. The authorities felt this highway would relieve the pressure on the inadequate, aging rail line. According to our records, an old road unfit for modern traffic already existed on that route, marking the way. Since both projects seemed urgent, the Highway Administration would have to send someone to survey the routes and inspect the local conditions to ascertain feasibility.

For three months I had been shoveling paperwork at the main office, growing bored and restless. Having had a long break after my work on the Lanzhou Highway, I yearned to get back into the field and do some real work again. I approached Zhao and volunteered to take over the reconnaissance for the Indochina-Yunnan Highway. He quickly accepted my offer. Thus I was able to leave the miserable city of Chongqing and hit the road again.

The first half of my journey led from Chongqing south to Guiyang, the capital of Guizhou Province. The road from Chongqing to Guizhou had been built by the provincial government long before the war without modern surveying techniques or engineering methods. It had numerous sharp curves and steep grades. The road had been planned to serve animal-drawn vehicles rather than modern traffic.

With shipments of military hardware and petroleum products from Indochina on the railroad to Yunnan increasing drastically, this trail between Kunming and Chongqing also had to be improved. Highway engineers were working frantically to alter the road's alignment so it could accommodate heavy trucks. However, the need for the road was so pressing that there was no time for careful surveys and the proper heavy construction. As a result, only part of this road was adequate.

I took a bus south. Traveling this road was a harrowing experience, to say the least, and we were grateful for the skill and experience of our bus driver as he successfully navigated the steep grades, narrow roadway, and sharp curves. He would step on the gas and ascend the grades at full throttle, then when the road had disappeared in front of the bus, he would make a sharp turn of nearly ninety degrees, jamming on the brake simultaneously to take the downgrade.

The driver negotiated hundreds of spots like this along the road. For hours we continued to climb up the first mountain range, and my barometer indicated we had gained two thousand feet in elevation by the time we reached the summit of the first pass. Then we began the gradual descent, which proved to be even more hair-raising than the climb had been. The driver took advantage of the steep grade by cutting off the motor and shifting into neutral, so the bus freewheeled down the slopes at breakneck speed. The fate of the thirty passengers rested on the brakes. We saw a wrecked bus in the bottom of the steep ravine; perhaps the gas the driver saved compensated for the lost bus.

When we finally crossed the range, which in Chinese was called the Great Trouble Ridge, we sighed with relief because a flat valley stretched before us. Before we had time to recover completely, though, we came upon another range of mountains. This range turned out to be even steeper than the last one, and crossing it was just as difficult.

For that entire day we heard only the coughing of the engine, the grinding of gears, and the squeal of the brakes. We passed such places as Hangman's Cliff and the Devil's Abode, as well as Great Trouble Ridge. As the names implied, others had felt the same way about traveling this route.

Not all of the river crossings along the way had bridges. Most of the wide, swift crossings were made on ferry barges built from heavy timbers. The barges were large enough to carry the weight of two fully loaded trucks.

The barges had no engines. The trucks were guided along two narrow planks from the wharf to the ferry. Twenty men with long, stout bamboo poles pulled the barge upstream for about half a mile, depending on the swiftness of the current. Then the barge was released, and it shot down the river like an arrow.

The crew members worked feverishly with their bamboo poles to steer the ferry across the river. The course was calculated to hit the other bank just at the point where the wharf lay. If the ferry hit the bank too early or missed the wharf, the crew would be unable to tie up. In that case, the swift current would carry the barge all the way downstream until it was stopped by a flat sandbar or smashed into the treacherous rocks that filled many parts of the river. The people on the wharf told stories of at least a dozen cases when barges were lost and all the cargo was destroyed. Only the strongest swimmers could save themselves from the eddies and currents.

Even a smooth crossing took at least half an hour. Since the barge could only carry two trucks at a time, heavy traffic would cause a fleet of trucks to back up on both sides of the river. Our bus was fortunate, because it reached most river crossings without having to wait. We covered the three hundred miles from Chongqing to Guiyang in only two and a half days, which was unusually fast. During my travels in the northwest, I thought I had seen the roughest terrain and the wildest rivers. Now I realized that what I had seen there were only mounds and ditches compared to this area. The mountainous slopes near Guiyang were so steep that virtually no level ground was suitable for farming. Western and northern Guizhou were covered primarily with yellow soils because of the great humidity, the relatively low temperatures, and the consequent incomplete oxidation of iron in the soil. Such soils tend to be expansive. The people of Guizhou characterized the soil as being "like the blade of a knife when dry and a horrible mess when wet." The popular description of life near Guizhou was that "the largest plot of arable land is three square feet; the people have nothing more than three cents worth of silver." These sayings were little exaggerated.

Guizhou Province had once been considered the poorest province in southwest China. However, with the increased wartime traffic from Kunming to Chongqing, the city had become an important stopover. Although millions were suffering from the effects of the war, one group managed to profit nicely: the professional motor-vehicle drivers.

The rackets the drivers worked were called "the yellow fish," "the aged wine," and "back-door goods." "Yellow fish" referred to extra passengers on chartered vehicles whose fare was pocketed by the individual driver. Since most of the vehicles on the road were uncovered trucks, and the passengers were packed together like sardines in a can, the term *yellow fish* arose.

The scam originated because of the insufficient number of vehicles and the fact that passengers would sometimes have to wait for weeks to get an available ticket to travel from one city to another. The drivers squeezed extra passengers into the already jammed truck; the passengers paid the driver more than the ticket-window price. If a driver could catch just two yellow fish on each trip, he made quite an income.

This racket was well-known to the authorities, who posted attendants at each terminal to verify the number of passengers on board. The drivers would simply tell the yellow fish to wait for them outside of town, where the drivers could pick them up undetected. The drivers would then drop the yellow fish off before entering the next town, telling the yellow fish they would come back for them after making the scheduled stop. Some particularly ruthless drivers would simply drive on without waiting for their fish. The poor victims were then left outside of town, with the money paid in advance for the ticket gone and their luggage sometimes left on the truck.

The big-time drivers did not bother with catching the yellow fish, because the "aged wine" scam brought them more money. This scam involved stealing gasoline from their vehicles. Gas was a precious, rationed commodity throughout the war, and posters along the roads read "One drop of gas is one drop of blood."

When experienced drivers were given new vehicles, their first move was to disable the odometer so the owners couldn't check the gas consumption. If the owners were strict and checked the mileage on a map, the drivers could still turn off the engine while going downhill. They could also tell the owners various tales, such as that they got stuck in the mud and extra gas had to be used to get out in low gear. The owners had the gas tanks sealed after each filling, but the drivers simply disconnected the fuel line at the carburetor and siphoned the gas from the line.

The owners tried adding dye to the gas so the source could be tracked, but black marketers were willing to buy the colored gas, so the racket continued. The common estimate was that for every ten gallons that went into the tank, one gallon went to the driver. The saying was, "No drivers are free of drinking the aged wine, just as no cooks are immune to squeezing the food money."

Carrying yellow fish and drinking aged wine did not satisfy the greed of the drivers, who were interested the most in "back-door goods." The government had imposed a strict ban on all foreign luxury commodities to clear precious cargo space for necessities. The drivers began smuggling French cosmetics, American cigarettes, Scotch whisky, and other expensive goods into the country. These goods were carefully hidden in the vehicles, and since the drivers were familiar with every detail of procedure at customs, they were able to bring the goods in unmolested. When they reached Chongqing, they sold the goods to the black marketers, who in turn sold them at unbelievably inflated prices to the few who could still afford them.

Before I left Guizhou, I witnessed a remarkable event. Normally, the Chinese want their parents to be buried in the traditional family cemeteries, which can be hundreds of miles from where they died. Given the rugged terrain of this province, only the rich could afford normal transportation.

A unique alternate business developed for transporting corpses. A magician-priest would accept a high fee to raise the dead and enable the corpse to walk under a secret spell. Grieving relatives would arrange to share costs with other families, so the magician usually raised several corpses at one time.

These corpses would allegedly follow the magician on foot over the mountains to their final destination. Then, under another spell, the corpses would collapse and would then be buried in their respective final resting places.

According to the magician-priest, it was essential that no living people come near the line of corpses. No reason for this mandate was given. As the magician led his unusual party, he loudly sounded his gong to alert people to his presence and the presence of the corpses.

Curious about this event, I was able to watch from a distance. The corpses were covered with black sheets, supposedly to hide their condition. They did not walk but hopped like rabbits close behind the magician.

Local residents told me that inns were located along the trail solely to accommodate such "guests." However, the business of raising the dead occurred less frequently as more roads were built and automobile traffic increased. I was fortunate to be one of the few to witness such an event; I offer no explanation.

After staying in Guiyang for a couple of days, I took another bus south to Kunming over terrain as rough as any I had seen during the first half of the trip. While en route to Kunming I had the opportunity to visit Huangguoshu Falls (Yellow Fruit Tree Falls). The falls are about 65 feet wide and have a drop of about 160 feet. Although the falls cannot compare to either Niagara or Victoria in size, I feel they are the grandest in China. The roaring of the falls could be heard miles away.

One spectacular feature of the falls was a temple of Buddha located behind the curtain of water. The temple was carved out of solid rock, and the entrance was hidden by the falls. Few tourists get to see this incredible sight. As the name implies, the area yielded delicious oranges, and a full basket could be purchased for very little money.

I reached Kunming safely in three days.

Chapter Ten

The Ho Chi Minh Trail

Kunming, the key city in wartime China, lay within an area that only five hundred years earlier had been considered completely barbarian. During the Ming Dynasty the emperor wanted to develop the southwestern territories, so he forced the rich people of Nanjing to emigrate to Yunnan Province, which was then a lightly populated, mountainous region. It must have been tragic for people whose families had lived in Nanjing for centuries to be moved forcibly to a remote hinterland.

Thousands died during this historic emigration, which poets described as the "exile to the ends of the earth." However, thanks to the high educational level of those who survived the journey, the emigrants rapidly developed the province. Even in the twentieth century, the accent of those who speak the Chinese language in Kunming is much like that of modern Nanjing residents, even though they are separated by several thousand miles.

Kunming stood at an elevation of six thousand feet, yet the weather was mild year-round. Because of its altitude it had no real summer season, but it enjoyed a very long spring leading directly into autumn; because of the southern latitude autumn never really turned into winter but lasted until spring. Like the people of southern California, the residents of Kunming were proud of their weather and boasted about their freedom from rain, snow, and even clouds over the city.

Before the war, Yunnan had been under the control of the provincial warlord, and virtually no social reforms had been instituted in the last decade. The worst problem for the people during those years had been opium. Yunnan's soil yielded an excellent crop of opium poppies, and the price was so low that those in Kunming who did not smoke the drug had become the exception. However, during the war, with the central government's full attention focused on local matters, people were no longer allowed to smoke.

When I arrived, I found that the people of Kunming rose fairly late. The stores on the main streets of the city rarely opened before noon, and one could buy anything well past midnight. Overall, I found the war had done more good than harm to Kunming. When the Ministry of Resources began developing the vast mineral resources in Yunnan, the production of tungsten and magnesium increased rapidly. As an expert on expansive soils, I found that Yunnan has the most explosive expansive soils in China, especially at Mengtzu near the Vietnam border.

Once I was in Kunming, I studied the best available map of Yunnan to find the route I was supposed to take. The first portion of the existing Yunnan-Indochina Highway ran fairly close to the railroad, as I had been informed in Chongqing, so I had little problem tracing the line. However, past the city of Kaiyuan the highway left the railroad. South of that point, little was marked on the map except contour lines indicating very rough terrain. The names of villages and towns the engineering office had reported to the Highway Administration did not appear on my map.

I later learned the reason for this discrepancy. No villages or towns existed along this part of the route, but the surveyors had needed some kind of notation to help identify specific locations on the line. As a result, they had simply made up their own names. If a particular location had red-colored earth, the engineers named the place "Red Soil Ridge." If they saw tigers nearby, they might name the place "Tiger Range." In fact, after the road was opened, some of the names remained unchanged and were eventually used by local people.

For five days I traveled south on a sedan chair carried by two men. Usually only foreigners and moviemakers were familiar with the elaborate chairs used by the mandarins. They had many comforts and mechanical conveniences and were carried by between six and sixteen men depending on the rank of the passenger. For the kind of territory I had to negotiate, a fancy chair would have been out of place. My sedan chair consisted of two bamboo poles with a rope net suspended between them. I crept into the net, and the men raised the poles to their shoulders. Once the poles were raised, I was pinned in the net like a mouse in a trap, unable to move any part of my body without effort. Very soon my neck would cramp, and I would be forced to get out and restore my circulation. I generally preferred walking.

Kaiyuan, where the headquarters of the road's engineering office was located, had become a wartime boomtown. All kinds of luxuries were available: canned milk, cocoa, French pastry, and Bordeaux wine from the colonialists in French Indochina. The engineer in charge of the headquarters was a man of considerable experience who had been working on the road for twenty years. He gave me a grand welcome and urged me to make my reports to Chongqing as detailed and accurate as possible so the officials there would get a true picture of the conditions out here.

"The bureaucrats in Chongqing know only one thing, cutting budgets," he said. "It is about time we got more money, considering the working conditions here."

This border town lacked the three basic necessities for the completion of a wartime road. The first need was good weather, but during half of the year this region experienced a tropical rainy season, and construction was impossible. The second necessity was acceptable topography, and this area was very rough, forcing the construction party to work in a dense jungle on rugged mountains. The third need was for harmony among all the people involved. Unfortunately for us, the provincial government did not wish to cooperate with engineers sent from Chongqing. With all of these influences working against us, I had little hope that the road would be finished on schedule.

The existing highway met the Yunnan-Indochina Railroad at Kaiyuan. The railroad was narrow-gauge and had been built by the French at the end of the nineteenth century to connect the Luizhou Peninsula with southwestern China. The railroad had been constructed with great difficulty and tremendous loss of life; the route was so mountainous that it required tunnels and bridges on nearly half its total length. At one point the French had considered abandoning the project. It took more than six years for the railroad to open to traffic, and the local people said a human life was paid for every tie that was laid. Our highway would face the same construction problems, and it was expected to be done in less than a year.

The engineering office in Kaiyuan sent two engineers to accompany me to study the road farther south. It was impossible to hire a sedan chair, even though they were in abundance in town, because when the carriers heard our destination, they refused to go at any price. I learned that the last group of sedan chair carriers hired by the office for the same destination met with tragedy. Of the twenty carriers who had gone, only five had returned; the others had been stricken with malaria and had died.

The only precautions travelers could take against malaria were to keep their shirts on during the day, sleep under mosquito nets at night, and avoid drinking the water without boiling it thoroughly. The sedan carriers, who were stripped to the waist under the hot sun, drank all the cool water they could find. As a result, very few survived a single trip.

We decided to start the trip on foot, accompanied by a few men who would help us cut our way through the jungle. Our provisions were carried by a few mules, and we were optimistic about a successful journey. The first day was not too bad. It was hot, and the trail was narrow, but we made steady progress, covering about twenty miles. At this rate I thought I could reach Hanoi in ten days.

The next day, however, I found the trail grew narrower and more difficult to follow with every step. The thick tropical growth covered the trail, and in places vines and bamboo entirely blocked it, so men with machetes cleared the way. We were told that within a day, or even overnight in the rainy season, the vines would grow over the trail again. Our progress became torturously slow as we hiked beneath the scorching sun and in the stinking smell of rotting vegetation.

Blood-sucking leeches were the most bothersome difficulty. As soon as they attached themselves to a human or an animal, they sucked the blood through a vacuumlike socket. The pain was usually slight and often was not even noticeable unless the leeches drained a large amount of blood. The biggest mistake we could make was to brush them away. When the body of the leech was struck, the socket snapped off, providing a rigid opening through which blood flowed continuously. The flow was very difficult to stop, since the skin could not be closed over the wound. The best trick was to stir the leech gently until it dropped off of its own accord. The leeches nested in tall, thick grasses, so we avoided such areas when we could. The first few men in our file were usually free from attack, but those at the rear of the line had to pass after the leeches had been disturbed, and they were struck more often. I saw a horse that had bled to death from an onslaught of these leeches, so I took the problem seriously.

As we continued southward, the jungle became even more dense. We saw so many species of tropical plants that only a botanist could have recognized them all. Vines sometimes extended for several hundred feet.

When night fell we built a bamboo platform supported by four bamboo posts, spanned with bamboo girders, and covered with dry leaves. Needless to say, spending the night on such a crude apparatus was fairly uncomfortable, but at least we were above all of the wild creatures that roamed the jungle floor at night.

We couldn't do anything about the snakes, which could slither up onto the platform with ease. I found myself unable to sleep, even though I was exhausted from the first day's hike. The jungle made strange noises at night. First I heard a soft strolling sound, as if a band of savages was slowly surrounding us, then there was a sharp shriek, as if a woman were crying for help. Even the insects and frogs sang strange songs in the jungle. Secretly, I wanted to get up and run or to yell back loudly at the voices, but I could hear the gentle snoring of my companions, who were used to these sounds. I finally drifted off to sleep and dreamed that I was being chased by a jungle cat, only to be saved by Tarzan swinging down from the treetops. When I woke up, I found everybody else was already packed and ready for breakfast.

The trail ran roughly parallel to the east bank of the Red River. Technically, the Ho Chi Minh Trail was an elaborate system of mountain and jungle trails used by the Vietcong to infiltrate south with supplies and troops into South Vietnam, Cambodia, and Laos during the war with the United States. Today both the Vietnamese and the Chinese refer to the trail that parallels the east bank of the Red River as the Ho Chi Minh Trail.

The bombing of North Vietnam increased during the war. Agent Orange was used by U.S. forces along the trail to kill the plants or inhibit their growth. Today we regard the chemical as a cause of illnesses suffered by veterans and their children. Agent Orange did kill the plants and animals along the trail, but it did nothing to discourage the continued use of the trail by the Vietcong. The vegetation along the trail may have been killed temporarily, but it will return in no time.

The Red River originates in the mountains of Yunnan and courses southward to its mouth at the Gulf of Tonkin. My orders from the Highway Administration included the instruction to study the Red River for possible navigation, which would save more than one hundred miles of roadwork. The river was wide enough, and the current looked smooth, so at first I thought the river would serve this purpose well. When we reached a stretch of river where local boats were available to take us downstream, I had my first real chance to see how the river was to navigate.

As we got into the boats I was surprised by their construction. Each boat was very narrow and had a bowed stern, similar to the sculls raced in the Olympics. The boat glided downstream on a smooth, pleasant current. Suddenly, I realized that our boat was nearing the edge of a waterfall. Before I had time to warn the crew, the boat darted down the shallow falls. I hung on tight to the railing, with water splashing all over me. A moment later we eased into another calm stretch of river, and the crew calmly bailed out water as though nothing had happened. Seeing my shaken and bewildered look, they explained that several dozen similar falls lay ahead and that the one we had just negotiated was minor. It seemed to me that if our flimsy boat hit one of the numerous rocks in the river while going through a falls or got tangled in one of the whirlpools below the falls, only the expert swimmers among us would survive.

I could see that the Red River could not be used for navigation. However, the three-hour boat ride was a thrilling experience. Even more important, we covered more distance in those three hours than we could have walked in three days.

We finally reached Hanoi after fourteen days of traveling along the Red River, nearly all of the time on foot. The French Indochinese immigration office required that all travelers have vaccination certificates for smallpox, typhoid, and other diseases. We did not have the proper certificates.

I went to the customs office, and in broken French I asked the officer there if there was any way around this requirement. "No," answered the officer soberly. "But I can introduce you to a doctor who will give you the required vaccination."

I took the hint and went to see the doctor. I found that for three times the regular vaccination fee, I could get the certificates for all of us without even rolling up my sleeve. I hurried back to my party.

We ran into another problem getting our baggage through customs, but the solution was similar. For a small fee, we could dispense with all of the pertinent regulations. If this was the common practice among French officers during wartime, I thought, the French have little chance of winning the war.

Vietnam had been under French control since 1884. The Red River Delta in the north and the Mekong Delta in the south were very productive regions. Rice production easily fed the people of Indochina, and large amounts had always been exported to China and other countries. The land is so rich that traditionally the people of Vietnam have not had to work very hard to survive, at least compared with most Asian peasants.

I had always found the French to be a lighthearted people, as opposed to the British or the Germans; the French did not take life as seriously. In France itself, however, I found the officials to be rather serious and strict regarding formal regulations. Here in the French colony, life was surprisingly different. The French colonials in Hanoi drank, feasted, and made themselves as comfortable as possible. I saw no sign in the entire city of the fact that France was at war. The people in Hanoi seemed unperturbed, even when the news arrived that Japan had declared war on France and that the nearby port of Haiphong was threatened by Japanese conquest.

I did not have time to do any further work. Word of the Japanese advance convinced me to leave after only three days in Hanoi. I returned to Kunming on the Yunnan-Indochina train.

The Japanese advanced rapidly and met little resistance as they captured Haiphong. When my train arrived at Kunming, I learned I had made a wise decision to leave Hanoi when I did, because the Chinese government ordered that one of the biggest bridges along the railroad behind me be demolished to check the advance of the Japanese and slow their possible invasion of southern China. However, the railroad was left paralyzed, and the highway project was abandoned.

– Chapter Eleven
Marriage and Leshan

When I returned to Kunming after a very tiring trip, I felt deeply disappointed. I was disappointed by the jungle of bureaucracy in Chongqing, by the Guomindang's halfhearted war effort against the Japanese, and by the widespread inflation and black marketing in free China under the Guomindang. The poverty and misery of the people in the northwest and the southwest pained me greatly, and I began to question whether the Communists were as worthless as all the propaganda claimed.

I was twenty-eight years old. For the first time, I began to think seriously about getting married and settling down. However, the right woman would not be a lucky person. In those days few sensible women would choose to marry a Chinese highway engineer and follow his difficult life.

The traditional Chinese marriages had been arranged by the parents of each prospective spouse, but modern Chinese, especially those who were better educated, no longer allowed their parents to make this decision. Even so, parents still played an important role in their son's or daughter's matrimony. I had seen tragedies result from blind marriages, but I had also seen the high divorce rate of the "love at first sight," American-style marriage. Since both approaches seemed to have failed in creating a happy family life, I thought there had to be a compromise between the two.

The famous Chinese philosopher Dr. Hu Shi was asked to compare the merits of the Western and Eastern systems of marriage. "Think of boiling water," Hu said. "In the Western tradition, boys and girls get married when their relations are at the boiling point. In the Eastern tradition, they start cold, then gradually come to a boil. As a result, the latter marriages last longer."

I believed most Americans chose their mates far too casually. When buying a house, one would carefully inspect the construction and consult with engineers and architects. It seems to me that as much care should be given to something as important as marriage. When Americans decide to get married, they rush off to the justice of the peace as if there were no tomorrow. No wonder one in three marriages in the United States ends in divorce.

I decided on a modified Chinese method: my parents would introduce a woman to me, but she and I would decide whether to marry. My parents agreed and introduced me to Edna Yu, a physics student at the University of Yangjing. After the fall of Beijing, the university had been moved to a small town called Dali, just north of the terminus of the Burma Road near Kunming.

In Dali she and I had the opportunity to spend a short time together and get acquainted. Dali was not exactly like the big city, and life on the wartime campus was less pleasant than that on the regular Yangjing campus in Beijing. However, Edna adapted to life in Dali. We both felt we would be able to adjust to the life of an engineer as well.

We were married January 19, 1941, in Chongqing. Among the guests at our wedding was Lin Sen, the nominal president of China. He remembered his close ties to my father during the revolutionary years against the Manzhou regime. As for the success of our marriage, in 1991 we celebrated our fiftieth wedding anniversary with our three grown children and four grandchildren. My theory about marriage is working for me at least.

Meanwhile, I had received orders to go to Leshan to work on the construction of the Sichuan-Xikang Highway. After the wedding I had to leave Chongqing at once. We decided to spend our honeymoon in Leshan, so Edna came with me. A truckload filled with people was leaving Chongqing for the city of Chengdu, and as honeymooners we were privileged to ride in the front seat next to the driver.

Chengdu, the provincial capital of Sichuan, lies in the center of the Sichuan Basin. Sichuan has long been described as a place "where heaven leaks." The climate is blessed by warm winters, early springs, long frost-free periods, and a long rainy season in the autumn. The high level of rainfall makes the province highly favorable for agriculture, and historically the area has also been called "the heavenly country."

The area gained this nickname after the building of the Dujiangyan irrigation system. In about 250 B.C., a man named Li Bing studied the Minjiang and organized the building of this irrigation system, which waters the Chengdu plain. It has enabled the area to enjoy fine rice harvests consistently, even during droughts. Although the system has been expanded over the millennia, Li Bing's initial scheme has been little improved. I believe it is the oldest irrigation system in the world that is still in use. Most of the great structures of the ancient world — including the Egyptian pyramids and the Great Wall — were built to glorify rulers, as sacrifices to gods, or for defense.

I have seen few structures built for the benefit of the people. In this light, Li Bing's accomplishment is notable.

My bride and I stayed in Chengdu for two days, celebrating our wedding with the finest food in all of China. Then we had to continue to our assigned destination in Leshan. We were fortunate to find a beautiful house that had been built by a Sichuan warlord.

For the first time in five years, I was able to find a comfortable place to settle down. Leshan was a beautiful city located on the banks of the Ming River not too far from Emei, the Sacred Mountain. We took time to climb this mountain using its countless paved steps and reached the Golden Top. The temperature at the top of Emei was close to freezing, but we passed the night at a guest house managed by the Buddhist monks who lived at the Golden Top, at an elevation of 10,140 feet.

When we rose, we hoped to see the "halo of Buddha," but we were disappointed, as it was a cloudy morning. The halo is a phenomenon that can be seen on still days when the sun comes out after the rain. Looking at the halo, the visitor can sometimes see his or her own shadow reflected in the misty air.

Edna and I did visit two other spectacular sights. One was the statue of the Buddha Puxian in the Temple of Ten Thousand Years. The temple was built in A.D. 980 during the Northern Sang Dynasty, and it houses a bronze statue of an elephant that weighs approximately sixty-two tons.

Next was the spectacular giant Buddha near Leshan. The Buddha was carved from solid rock out of a mountainside. The figure is 230 feet high, with shoulders 92 feet wide and fingers over 27 feet long. According to legend, a Tang Dynasty monk was concerned that this spot was a hazard to boatmen and wanted the Buddha there to protect them. Work was begun on the statue in A.D. 713 and was finished ninety-eight years later. The Buddha is almost four times higher than the presidents' heads on Mount Rushmore.

The Leshan-Xichang Highway represented one section of the Sichuan-Xikang Highway. My assignment was to experiment with a new type of pavement that would utilize local materials at a lower cost and with a better service life than previous pavements. At the same time, I was to train four newly arrived graduates of the University of Iowa. This assignment was comparatively easy, so I had time to socialize with various college professors at the two prominent universities, which had been moved from occupied areas to Leshan.

The first section of the Sichuan-Xikang Highway was completed a short time after Edna and I had settled down in Leshan. Because of the importance and extremely difficult construction of this road, the commissioner of highways, Zhao Zukang, came to Leshan to lead the first vehicles over the road. I was asked to join the inspection party.

Marriage on January 19, 1941, author and wife, Edna Yu

Golden wedding anniversary

The Devil's Abode, a treacherous mountain road

This was the first time I saw what my fellow engineers had accomplished. After my previous experiences, I thought I had seen difficult construction and terrible terrain. I changed my mind when I saw Devil's Abode on this route. Devil's Abode was a three-sided canyon in which the road had been carved into nearly vertical canyon walls of solid rock, looping along the three sides of the canyon. U.S. engineers would no doubt have wanted to build a high bridge spanning the canyon, but that would have required piers at least three hundred feet high, the cost of which we couldn't even contemplate at that time. Cutting the road into the nearly vertical walls had been dangerous enough. Many stoneworkers had slipped and fallen into the canyon from a height of over three hundred feet.

We got out of the trucks and walked the length of the canyon. Zhao stopped for three minutes of silence in memory of the 125 men killed during the project. It was heartrending to see women kneeling on the edge of the road, lamenting their dead husbands. The compensation they received was very small, certainly not enough to cover the livelihood of the survivors. The highways built in China during the Sino-Japanese War were truly built with blood and tears.

Upon returning to Leshan, I found a telegram from the Highway Administration in Chongqing waiting for me. It read, "You have been appointed assistant chief engineer for the Burma Road. Report to Xiaguan immediately. Take as many engineers with you as possible."

This took me completely by surprise. Appointing a man who was only twenty-nine to such a responsible job was unheard of in China. Everyone realized that the Burma Road was now considered to be wartime China's lifeline, the most important construction job during the Sino-Japanese War. It was a great honor to become a part of it.

Everyone in my crew was excited at the news, since they all realized the importance of this challenge. We left only the accountant, who had to wind up unfinished business; all the engineers were ready to go, including the four graduates of the University of Iowa. They had heard a great deal about the Burma Road in the United States. Our group also included Li Wenping, who had been my schoolmate at the University of Michigan.

It took us only a day to pack. In three fully loaded trucks, we set off for Xiaguan, the headquarters of the Burma Road administration. I was welcomed warmly by the engineers who were already there, and the first thing they wanted to do was challenge me to a game of table tennis. My reputation as a good player was well-known. My wife was excited about our new assignment, because Xiaguan was only a short distance from Dali, where she had gone to school. The most important work I had done yet was about to begin.

– Chapter Twelve
The Burma Road

The landmass of China reaches its highest altitude in the west near the Tibetan plateau, nicknamed the "Roof of the World." The continent slopes downward to the east. This slope consists of three sections divided by two steep steppes. The higher steppe consists of the Kunlun Mountains and roughly follows a ten-thousand-foot contour line. The second steppe follows the Greater Khingan range, the Taihangshan, the Wushan, and the eastern border of the Yunnan-Guizhou plateau at an elevation of about thirty-five hundred feet. The major drainage system consists of the Yangzie (Yangtze) and the Huanghe (Yellow River) and follows the general slope of the overall landmass from west to east, emptying into the Pacific.

The exception to this general west-to-east downward slope lies in western Yunnan. In that region parallel chains of high, rugged mountains known as the Hengduan Mountains, or the Horizontal Cut, range in altitude from twelve thousand feet in the north to less than six thousand feet at the southern end. The drop from the peaks to rivers in the valley floors often exceeds six thousand feet. Where river valleys are the narrowest, two people standing on opposite summits can almost hear one another shout, but to visit face-to-face they would have to spend a day climbing down one scarp and up the other. The river currents are so swift they are unnavigable.

The total length of the Burma Road, which starts at Kunming and ends at the border between China and Burma, was to be 717 miles, approximately the distance from Kansas City, Kansas, to Denver, Colorado. It was not a long highway compared with other roads I had built during the war, but it was one of the most difficult. The deadline and the rough terrain, coupled with the hostile wartime environment, compounded our problems.

The alignment of the Burma Road was such that we had to cross the Horizontal Cut Mountains. The course of the road crossed three major mountain ranges and two rivers, the Mekong and the Salween. The road reached an elevation of eighty-five hundred feet, then dropped sharply to three thousand feet across the Mekong. The route rose sharply again to nearly eight thousand feet, only to drop again to the Salween at an elevation of three thousand feet. One way to visualize this was to imagine a road crossing the Grand Canyon — twice — although the Grand Canyon, which is about a mile deep, is shallower.

Hairpin curves would be necessary to gain five thousand feet within the short distances we had to negotiate. We constructed as many as thirty-five hairpin curves going up and the same number coming back down. We rarely saw a stretch of flat land more than a mile long, with the exception of a short stretch near Kunming and the town of Baoshan.

With more time our surveying party might have been able to select a better route, but our deadline was too close. Our construction crews followed right on the heels of the surveyors, and we had no margin for error. As soon as the surveyors staked the line, it became final. In the rough terrain, if we could limit the necessary excavation of earth to thirty-five thousand cubic yards per mile, we considered it good engineering. We could have saved a great deal of distance and excavation if we had used tunnels, but tunnels create bottlenecks in the construction line, and our schedule would not allow the delays.

Of the thirty million cubic yards of earth to be moved, more than forty-two percent was classified as rock excavation. All of this work was done by hand with only picks, shovels, and bamboo baskets. At the peak of construction we employed fifty thousand laborers. When we lined them up along the roadway there would be one person every five feet.

Simply organizing that many people in one place at one time and trying to increase their efficiency were major tasks for the engineers. None of the workers had previous experience in road construction, and when they began each worker was able to handle less than one cubic yard of earth daily. As they learned to do the work, their productivity increased to as much as four cubic yards in a normal twelve-hour day. About one-fifth of the workforce was female, and they did the job as well as the men.

Rock excavation required the most time and skill. We drilled blast holes with a star drill and an eight-pound hammer. In this manner, a team of four could drill two three-foot holes in a day. The holes were then charged with either dynamite or blasting powder and detonated. Workers hauled away the rock debris in bamboo baskets. Our crews moved more than one million cubic yards of solid rock in this fashion.

Our worst enemy was the rainy season. It began in northern Yunnan as early as June and lasted until late September. The local residents stopped all activity during the rainy season, but we did not have this luxury. Construction went on at a much slower pace than before, and to work in those conditions engendered sickness and death.

Watching the rainfall all day long, drizzling every day with no respite, became tiresome and depressing. We had poor rainwear, so our clothes quickly became soaked each day. The death toll among the workers began to mount from both disease and accidents, eventually reaching five percent of our workforce.

We divided the road into seven engineering divisions, and each was subdivided into four or five sections. The crew in each section was equipped with a radio and a telephone and included radio technicians, doctors, blacksmiths to repair tools, and teams of women in charge of knitting straw sandals and bamboo baskets. Each section crew could handle all of its own problems without a decision from the division manager. This became one of the best-organized projects I ever saw. Best of all, I had returned to the real work, as opposed to putting endless chops on paper in Chongqing.

Soon after the road opened to traffic, we began to experience landslides. We had cut the side slopes very steep because of our severe time constraints, and the incessant rain had saturated the soil. Slides occurred with alarming magnitude and frequency, sometimes several at once. After a particularly severe rainstorm, over fifty slides blocked the road.

Most of the slides were not serious and could be cleaned up quickly, allowing traffic to resume. The serious slides involved base failure. When the decomposed granite and the lateritic topsoil became saturated, they began to move down the slope, and the entire mountain would seem to be shifting.

We tried various methods of slide prevention. The crews dug ditches at the crest of slopes in potential slide areas, but this accomplished nothing. Retaining walls, supported with long pilings driven into supposedly stable ground, were constructed at the toes of slopes, but these walls disappeared overnight under the next slide. Finally, we even went to the extreme of building masonry arches over sections of the roadway where we expected slides, so that sliding mud would cover up the arch and form a tunnel. This succeeded, but it was terribly expensive. The least expensive way of handling the slides remained simply to clear each one by hand. After one major slide we had a thousand workers clearing the road.

During this time I heard about a novel solution for travelers facing a slide. A dignitary driving from Rangoon to Kunming on the Burma Road was stopped by a huge slide. At the same time, a Chinese officer headed for Rangoon had to halt on the opposite side of the slide. They hiked over the mud and agreed to trade cars, so both could turn the vehicles around and reach their respective destinations.

I advised that we locate another route and completely avoid the least stable area. This would require time and would increase the length of the road, but it had to be done. When we implemented this decision, we avoided the worst of the slides completely.

Traffic on the Burma Road soon reached more than three thousand vehicles per day, of which more than ninety-five percent were heavily loaded trucks. We realized that the gravel surface was inadequate for the traffic. I was assigned to pave the western section of the road with blacktop.

The government in Chongqing had made a deal with Texaco Oil to supply all the asphalt we required. As a stipulation of the agreement, Texaco demanded that the company be allowed to appoint an engineer to oversee the construction of the new surface. Texaco sent a young man fresh out of college who had no field experience. He was armed with a copy of the manual *Asphalt Pavement Construction,* published by the Asphalt Institute, and he duly reported to his superiors in Washington.

At the same time, on advice from the U.S. government, the Highway Administration in Xiaguan decided to mechanize the entire paving operation. Aggregate crushers, diesel rollers, and dump trucks were on their way. With all that equipment, we expected the paving job would be done in record time.

During this period military news from the Burma front was completely discouraging. U.S. Lieutenant General Joseph W. Stilwell was losing ground as the Japanese advanced steadily. The millions of tons of war material stored in Burma had to be moved before it fell into the hands of the Japanese.

Generalissimo Jiang Jieshi personally came south to visit the road. He gave orders that the road had to be paved at all costs and within the assigned time. Since the entire Burma Road operation was under military jurisdiction, and I was given a military rank, we had no alternative but to follow orders.

I was promoted to chief engineer and given complete charge of the paving work. Feeling the weight of this responsibility, I began to work day and night to organize the operation. However, nothing seemed to fall into place.

The aggregate crusher we had ordered worked beautifully for several days, but the purchasing agent had not ordered spare jaws. The existing set of jaws wore out within a few days, and spares could not be shipped for several months. The diesel-powered rollers broke down, and our mechanics could not fix them. In desperation I returned to traditional methods and mobilized an army of laborers, both men and women, to crush and break the stone by hand. Once again crews pulled the stone rollers by hand as well.

The Highway Administration could not understand why, with all the money invested in mechanical equipment, the equipment lay idle. I was severely reprimanded by the administration for not taking good care of the equipment. They simply did not understand that without supporting facilities and spare parts, the equipment could not function for long.

That was not my major problem, though; the Texaco engineer filled that role. He insisted that the road be paved according to specifications outlined by the Asphalt Institute and threatened to stop shipments of asphalt if the specifications were not rigorously followed. Unfortunately, this agreement was in the contract the government had made with Texaco.

Luckily, the Texaco engineer also adhered strictly to his own working hours, which followed American practice: eight A.M. to five P.M. with an hour lunch break. Consequently, when he was on the job we did whatever he demanded. After he left, we worked through the evening hours and paved the road as fast as we could. In this fashion we managed to work around our American adviser and finished the paving job within a reasonable time.

I knew well that the pavement I put down was not in accordance with the Asphalt Institute's specifications, and it would not have the life expectancy they indicated. However, I expected the use of this road to be short-lived. The pavement we laid proved to be adequate.

On December 7, 1941, my crew was working out of Wanting. In the evening we heard on the radio that Japan had attacked Pearl Harbor, and the United States had entered the war against Japan. We were elated, because China had been fighting effectively for six years without allies, since France and Britain had to give priority to their war at home over protecting their Asian colonies. Now China had the most powerful ally of all.

China had lost all of its eastern areas, and millions of Chinese had died, yet the world had still considered this a local Asian conflict. Now at last we had an ally capable of bringing Japan to its knees. We broke open a bottle of whiskey we had been saving for years for such an occasion.

I know it was a sad day in the United States, but it was one of the most joyous days of my life. Finally I could foresee the day I would be able to go back to Nanjing and Beijing. The hardships I had endured during the past six years might come to an end.

On December 29, China became a full-fledged ally of the United States, with Generalissimo Jiang Jieshi named the supreme Allied commander for China. Consequently, the traffic on the Burma Road increased even more. I managed to push the blacktop work to an area near Baoshan, and by that time I had convinced the Texaco engineer of the gravity of the situation. Our relations improved, and he became much more cooperative.

By May 1942, the news from Burma had become dismal. The merchants and shopkeepers in Lachu had realized the Japanese would not be stopped

*Hairpins on
the Burma
Road*

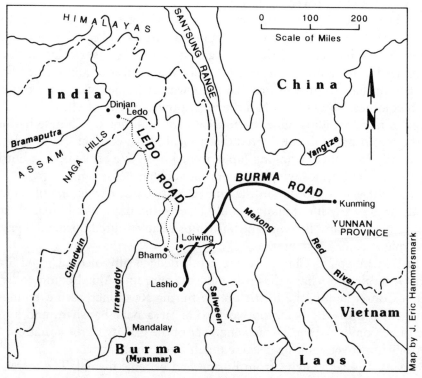

The Burma and Ledo Roads (map by J. Eric Hammersmark)

The first use of mechanical equipment in road construction in China
(left) Dump truck (right) Mechanical roller

Women preparing aggregates by hand (using hammers)
for use in pavement construction

Application of asphalt by hand

from reaching their town. Those who could leave abandoned all of their merchandise and most of their personal belongings and joined the bumper-to-bumper traffic along the road to Kunming.

I saw that more engineering work on the road would be a wasted effort, because survival had become the order of the day. In the growing chaos I radioed the Highway Administration at Xiaguan for instructions. The director was in Kunming for dental work, so no one in the office could make critical decisions. Finally, I received a somewhat evasive reply instructing me to evacuate all minor personnel and to take a skeleton crew to Wanting for orders from General Yu.

Six months earlier, my daughter, Dorothy, had been born. My wife and daughter were the only family left in the area and were still living in the construction shed. I did not think the situation was as serious as it appeared. At noon one day a station wagon arrived, and my friend who was in charge of telecommunications was surprised to see my family still at the post. "You're crazy to have your family here," he said. He had just driven up the road from the China-Burma border.

I decided to have my wife and our baby leave with him in his car. It took us fifteen minutes to pack them up and send them away. I thought it might be that last time I would see my family.

In the United States, the public story of the Burma Road is somewhat inaccurate. Most Americans think the road was built by U.S. engineers. Actually, there were no Americans on the Burma Road during these phases of construction with the exception of the lone Texaco engineer. However, the Burma Road met the Ledo Road at Mongyu, Burma, and the Ledo Road continued the route west. The Ledo Road was built by American engineers.

Construction on the Ledo Road, which is five hundred miles long, began in 1942. The project employed U.S. Army engineers, most of whom were African Americans, and thirty-five thousand local Burmese workers. The two linked roads became an overland supply route across Burma from India to China. One particularly important heavy cargo was airplane fuel for bombers and fighters. However, the $150 million Ledo Road, also known as the Stilwell Road, was not finished until 1945 when the war was almost over.

– Chapter Thirteen
Crossing the Salween

When I faced the Japanese soldiers at the top of the ravine on the Burma Road, the instant I heard the rattle of the machine gun I could see the faces of my mother, my brother, my sister, my wife, and all of my friends flash through my mind. At the same instant, instinctively, I fell backward, dropping abruptly over the edge of the sloping side of the ravine.

I tumbled about fifty feet down the slope and was unconscious for a few minutes. When I came around I heard the rattle of the machine gun again, and more bodies dropped into the ravine. As the shooting continued, I dared not stir, even though my leg hurt a great deal.

I lay in the bottom of the ravine for what seemed like hours. Slowly, the jungle around me became dead quiet. When the shooting had stopped for some time and the heavy boots of the Japanese no longer hammered the ground, I finally got to my feet and looked around.

The first sight to greet me was the body of a fellow engineer, which was hanging in a tree. When I touched him, he was cold. His name was Wu Junwei. He came from Guangzhou and was a brilliant young man not more than twenty-five years old.

I walked aimlessly through the wilderness, trying to find my way out of the ravine without any bearings. Late in the afternoon I came upon a dozen other refugees. I knew a couple of them, and we sat down to map our strategy. We decided that the first priority was to find something to eat, since none of us had eaten anything for at least two days.

Eventually, we spotted a lone farmhouse. A Japanese flag flew from the top of the roof, so apparently the Japanese had used the place for temporary quarters; however, we saw no one now. Upon venturing closer, we found the place was empty. The owner had either fled or been killed. I was tempted to take the flag as a souvenir, but the others stopped me, pointing out that doing so might attract the attention of the Japanese. I reluctantly agreed.

We searched the small house and found some rice in the kitchen. Still worried that the Japanese troops might return, we took the rice and a cooking pot and left quickly. We moved through the jungle wilderness toward the Salween.

Late that day we finally reached the riverbank. We estimated that we were at least ten miles west of the remains of the Salween Bridge, where we believed the Japanese were concentrated. At that distance we felt safe gathering some branches and starting a cookfire. When the rice was cooked, we ate with our hands because we had no utensils, but at least we had full bellies.

The headwaters of the Salween are in eastern Tibet. Spring rains had raised the river level more than fifty feet, and vessels could not navigate most of the river because it was so rough. Now, in May, the water level had reached its peak, and consequently the river had become over a mile wide at our location. Local Burmese living along the upper Salween used bamboo rafts for transportation, but if the rafts reached the lower stretches of the river, which were filled with rapids, they quickly fell apart. From the bank we saw lots of loose bamboo floating in the river.

Since the bridge had been blown and we had no way of making a raft, we could only cross the river by swimming. I felt I could do it, and I asked the members of our group if any strong swimmers wanted to come with me. Only one person accepted the challenge. He was a champion swimmer who at one time had swum across the Huangpu River at Shanghai.

He and I decided to try the crossing. I warned him that the current was very swift and that it was necessary for us to reach the north bank as fast as possible. Otherwise, the current would carry us downstream right into Japanese-occupied territory.

One of my friends, a nonswimmer, insisted that we take him with us. Someone found a large bamboo pole. We tied him to the middle of the pole, and the champion swimmer took the head of the pole. I held the other end. Everyone else just looked at us and shook their heads.

We plunged into the river and swam as fast as we could, although the current carried us downstream as we angled across it. When we reached the middle of the river, Chinese soldiers on the north bank opened fire on us, apparently believing we were Japanese soldiers. We shouted as loud as possible that we were Chinese refugees. They withheld their fire, but the current carried us downstream past them.

Farther downstream a Japanese patrol spotted us and thought we were Chinese soldiers; they opened fire on us, too. Luckily, none of the bullets came near us. We were probably out of their range. Still, we did not dare slacken our pace.

We finally reached the north bank and climbed out of the water. I laid down on the bank, too weak to move. Soon, however, my nonswimming friend urged us to proceed. We found a rugged trail that led in the direction of Baoshan, which we judged to be about sixty miles north. When night came, we were so exhausted that we did not even try to select a good place to sleep. We just laid down where we were and slept the sleep of the dead.

At daybreak we were cold, wet, and hungry once again. However, we headed for Baoshan, which was now only one long day's hike away. At least the Japanese remained on the far side of the river behind us.

We found Baoshan, the second-largest city in Yunnan after Kunming, to be a dead city. Corpses lay everywhere, and disabled trucks jammed the streets. The survivors had evacuated the entire city, and the panic had probably caused the violence; we knew the Japanese had not yet crossed the Salween. In any case, we could not find any food.

However, I was pleasantly surprised to find my colleague, Dr. Zhu Guoxi. He had left the construction shed at the side of the Burma Road in the pickup truck with the rest of us, but we had lost each other when we abandoned the truck. Zhu, who was from the Imperial College in England, had been the head engineer for the construction of the Salween Bridge. When I saw him I thought I was dreaming. It turned out that he had been fortunate enough to miss the Japanese and had walked along the Salween until he came to the walled city Tengchong. Not far from the city he had crossed the river by means of a rope bridge, and he reached Baoshan at the same time I did.

All of us in our small group agreed we had no reason to stay in Baoshan. We started for Xiaguan. Fortunately, before long I spotted a truck that was nearly full of people and was heading the same direction. We reached the truck as it slowed down for a steep grade. Grabbing the tailgate, I hoisted myself into the truck, despite the protests of everyone already onboard.

I got off when the truck reached Yongping, a small town that contained the headquarters of the fifth division of the Burma Road engineering force. I hurried to the headquarters on foot. When I appeared before the division chief, he acted as though he were seeing a ghost. "Everyone thought you were dead," he said.

I radioed my wife in Xiaguan, telling her that I was very much alive. The division chief arranged for a hot bath, after which I threw away all of my clothes and replaced them with new ones. Then I had a hearty meal, the first decent meal I had eaten in a week. When I finally went to bed, I slept for twelve hours.

When I awoke I said to myself, "I am reborn." I had seen death staring at me down the barrel of a gun, and I had hiked, swum, and hiked again to get back here. Now I had a second chance to live.

*Author with wife, Edna, and daughter, Dorothy, after escape from
the Japanese on the Burma Road, 1942*

I wasted no time heading for Xiaguan. My wife and I had a tearful reunion
after a ten-day separation, but the uncertainty and violence made us feel as
though we were meeting in another world. She told me everyone had assumed
I was dead. Only her faith in God had kept her hope alive. Of course, my six-
month-old daughter knew nothing about my adventures. She remained in
perfect health.

At the Highway Administration center, the engineers staged a rally
attended by more than three hundred people. As a survivor of the recent chaos
on the Burma Road, I was asked to speak. The central theme of the rally was
to remember those who had died during the recent Japanese advance.

The underlying motive, however, was to protest the way the Highway
Administration had handled the entire incident. The victims had died for no
reason because of the failure of the Chinese army to stop the enemy and the
lack of orders for us to withdraw. A huge poster was erected in front of the
platform that read, "The eyes of the dead remain open," implying that the
ghosts of the victims would seek revenge.

The mother of Wu, who had died next to me at the ravine, attended the rally. Wu had been her only son, and she had flown from Guangzhou just to attend this rally.

I was filled with shame over my role in the retreat. The entire rally felt like a knife being stabbed into my heart. With hindsight, I knew I should not have been so timid when I had first called on General Yu. I should have demanded clear-cut instructions, but I had not done so. When I failed to hear from the general, I should have ordered the evacuation at least twenty-four hours sooner. If I had, perhaps all of my engineers would have survived.

After the rally I decided I had no reason to remain on the Burma Road. My wife and I went to Kunming for a few days of rest.

Zhu, myself, and my fellow engineer, all having survived the Burma Road retreat, got together for a reunion and a memorial picture. I had the opportunity to meet Zhu fifty years later in Shanghai. He was well, and his son was following in his father's footsteps and was working on his Ph.D. in engineering. It pleases me that today a monument stands at the entrance of the Burma Road in memory of those who died there.

– Chapter Fourteen
Roof of the World

After the loss of the Burma Road, talk at the Highway Administration in Chongqing turned to development of the northwest. The need for an overland route was centered on two more proposed roads, the Qinghai-Tibet and Xikang-Tibet Highways. Up until now, the people of Chongqing had only known of Tibet as a place on the map.

The first phase of the Qinghai-Tibet Highway was constructed entirely within Qinghai Province. It was to terminate at Yushu, a total length of 570 miles, where it would meet the Xikang-Tibet Highway. Qinghai, known on Western maps at the time as Kokonor, was sparsely populated. The ethnic composition of the modest population was diverse. The greatest concentration of people lay in the Xining area. Tibetans were scattered throughout the province. The Mongol communities were clustered especially in the Qaidam vicinity. A large community of Tu people had developed in the country north of Xining; the Hui (Muslims) lived primarily to the southeast. Khazak villages lay to the west of Qaidam.

Even with the remarkable ethnic diversity of the population, the various groups seemed to coexist well. In Xining I noticed a Muslim restaurant where no pork was allowed, and a Han eatery stood across the street that served all the pork anyone could want. The fact that the people lived peacefully here stood in contrast to the religious hostility in the Middle East and the Indian subcontinent.

One exception to the harmony occurred when our survey party met the Ngolog tribe. Some of them were cannibals. They took great pleasure in killing people and hanging the skulls on top of their tents as trophies. We came into their village ignorant of their ways, and they detained us.

No one in our group could speak a word of the Ngolog's language. They refused to release us, and we did not know what their intention was. Then one day the head of the tribe tried out one of our rifles. He fired several shots at a distant target and hit only once. Our captain of the guard took a turn and scored every shot with the same rifle.

The chief was deeply impressed. He not only allowed us to go free but gave us some dried meat as a token of respect. We hoped it was not human flesh.

In summer 1943 Zheng Yangfu, the minister of communication, appointed me chief engineer of the Qinghai-Tibet Highway. He interviewed me briefly and finished with four brief words: "succeed, or don't return."

I took a large crew of engineers and laborers through the mountains to Xining, the provincial capital of Qinghai, where I called on the governor to gain his confidence and cooperation. Governor Ma was a devout Chinese Muslim, called a Hui, who complied strictly with Islamic laws. I made sure he knew I respected his religion.

To begin our survey we needed one hundred saddle horses, forty tents, at least fifty yaks, and a number of sheep. Governor Ma supplied them and also placed twenty guards at our disposal. These were all essential, because as soon as we left Xining we were in no-man's-land, climbing farther up the Tibetan plateau. Xining stood at an elevation of 7,680 feet, and by the time we reached the Sun and Moon Mountain we would have climbed to 11,800 feet.

Between Huangyuan and the Sun and Moon Mountain we encountered no one. To the south of the mountain we could see a vast desert. I recognized the area as the setting of a famous story from the Tang Dynasty about Princess Wen Cheng.

In A.D. 198 Princess Wen Cheng, a daughter of the royal house of Emperor Tai Zhong of the Tang Dynasty, came to the mountain. She was betrothed to Sangzan Ganbu, the ruler of Tufan, which is now Tibet. She brought advanced techniques from the Han people for making pottery, paper, and wine and thus made great contributions to the economic and cultural life of the country.

Supposedly, when Princess Wen Cheng arrived at the Sun and Moon Mountain, she looked to the north and saw green valleys, but to the south she saw nothing but the desert and refused to go on. A messenger took word of her refusal back to her father. Because of his daughter's resistance, Emperor Tai Zhong tried to negotiate an annulment of the marriage, but Sangzan Ganbu demanded a payment in gold the size of the sun and the moon. The emperor refused, so the princess continued from the mountain on to Tibet. This event gave the "Sun and Moon" Mountain its name.

Legend also says that when Wen Cheng traveled from Changan to Lhasa, she carried a mirror in which she could see scenes of her homeland. On the journey she broke the mirror, which turned into the mountain of the sun and moon. When the mirror broke, the princess gave up thinking about her home forever.

Princess Wen Cheng made a great contribution to Tibet. In her memory the Tibetans built a temple with her image carved in stone. A folk song of welcome to her has been handed down through the generations to the present day:

> On the fifteenth day of the first month,
> When the Princess consents to come to Tibet,
> No one is afraid of the vast sandbars,
> And there are a hundred steeds to meet you,
> No one is afraid of the snow-capped mountains,
> And there are a hundred dzos to greet you.
> No one is afraid of the surging rivers,
> And there are a hundred cowhide boats to welcome you.

Life on the plateau was dramatically different from that in any of the heavily populated areas of China. With an average elevation of more than eleven thousand feet, sudden changes in the weather posed a major hazard. No real summer ever came to this plateau. Even during the warm season in July and August, a snowstorm could arise suddenly. Even worse than the snowstorms were the hailstorms, with hailstones the size of a chicken egg. We found no shelter on the high plateau other than our flimsy tents, which could be punctured by the hail.

A local saying was that in "April, May, and June, the rain and hail can bring you to tears." In the months after August, if travelers still dared to move through this region, they faced a constant risk of frostbite. Without the proper equipment, it was easy to lose an ear or a finger.

Our main food was roasted barley flour in a form called zanba. The flour was mixed with butter and sugar when available and was then kneaded into a ball. It required no cooking, and we ate it on horseback. The zanba wasn't bad, but a constant diet of it caused painful constipation. For meat, we relied mainly on our shooting skills. Wild goats were plentiful in the grasslands, and each goat provided food for three people for several days.

Hot liquids were even more critical than food on the plateau. The typical Tibetan drink was made from tea bricks imported from India. One tea brick boiled in a kettle and topped with butter could warm the stomach for most of the day. No one who traveled the plateau could go without tea to help ward off the bitter cold.

Cooking presented a problem, since the average elevation was above the timberline. Without trees we had no firewood. We collected wild yak dung for fuel, which burned very well when dry but was useless when wet. We gathered as much yak dung as possible during dry weather and stored it in our saddlebags so we could look forward to a hot meal. The left saddlebag was for yak dung, and the right was for wild goat meat. At times, when the rain continued for several days, we could not build a fire, so we were forced to eat the goat meat raw. It wasn't really bad when it was frozen. Another local saying was that to survive in the grasslands of the plateau, three things were needed: a fast gun, a swift horse, and a sheepskin coat.

The origin of the Qinghai-Tibetan plateau was a focal point of interest for geologists, who considered it a possible keystone in the science of matter and motion in the earth's crust. The glaciers of the plateau act as reservoirs of frozen water and are the fountainheads of the mighty rivers to the east. The Yangziejiang (Yangzie) emerges from the glacier of snowcapped Mount Geladandong and Mount Yarlungzangbo, and the Jiemayangzong glacier. The roaring torrents of ice and recently thawed waters also converge to form the Huanghe, or Yellow River.

With water in such abundance, one might think travelers would have no problem finding drinking water. On the contrary, in the desert north of the Kunlun Mountains, water supplies were sometimes miles apart. Occasionally, we came upon stagnant water in a gully, but these pools were usually polluted with animal droppings. One way to obtain drinking water in the grasslands was to dig a pit and let it fill from the seepage of groundwater. The water in the pit turned bitter after several days and was no longer fit to drink.

We encountered some unusual features on the plateau that we had to consider in surveying the road. These features were marshy meadows, locally known as *ju ru* in Chinese or *na keng* in Tibetan. The marshy meadows had high and low spots, like bumps and swirls on a beehive. The high spots were grassy, and the lows were filled with water. Travelers were warned not to cross these areas and were advised to go around them.

In sandy soil the water accumulated in the low areas, but it was only a few inches deep. During the spring months the water rose and flooded the entire area, giving the appearance of a vast lake. These marshes were generally above fifteen thousand feet, near the Tanggulia Mountain range. They could be drained with good engineering.

Another type of marshy meadow was similar in appearance, but the upper soils consisted of decomposed vegetation, and the subsoil was fine-grained mud. These formations could be more than thirty feet deep. Our road had to cross such an area at Chelaping, at an elevation of nearly 16,400 feet. The area was several hundred square miles in extent, with solid ice formations underlying the marsh only a few feet below the surface even during the summer months. Riding a horse across these marshes gave me the feeling of crossing an elastic rubber surface. The vibration of the horse's hooves could be felt hundreds of feet away, like tapping on a great bowl of gelatin.

On the plateau, legend and contemporary scientific work have come to an apparent agreement. Legend claims that the Qinghai-Tibetan plateau was once a huge sea until five evil dragons sucked the water dry and changed the area into mountains. Geologists have found two fossils of ichthyosaurs in the plateau, which are animals that lived in the sea one hundred million years ago and fed on mollusks.

Some of the marshy meadows we crossed had been formed millions of years ago. The area had no source of flowing water but retained subsurface pools of icy water. We had to exercise extreme caution when riding across these marshes, since one step into the marsh could cause a horse to lose its footing and sink too deeply to get out. In the area of Yematan, or Wild Horse plateau, we saw the remains of wild yaks, sheep, and horses that had become stuck in the marsh and had died and rotted.

Not all of our travels on the high plateau were miserable. We came to a hot spring about a hundred miles north of our Yellow River crossing. The spring was boiling hot, but it was immediately adjacent to another spring of cold water. By diverting the cold water into the hot spring, we adjusted the temperature to our liking. We pitched a tent over the hot spring, and everyone enjoyed a very pleasant bath that washed away our stress from crossing the plateau.

Written reports of the geothermal activity on the plateau date from the late nineteenth century. However, reports of active volcanoes in northern Tibet proved to be false. Recent investigations have located no fewer than six hundred areas in Tibet that produce either steam or hot water above the ground. Regarding Qinghai, the hot spring we visited is probably the only one known.

Our advance party took sixty-five days to travel the 520 miles to Yushu, our eventual rendezvous point with the Xikang-Tibet Highway. Now that the route had been surveyed, it was ready for construction. We returned to Xining to begin the real work.

The construction of the Qinghai-Tibet Highway actually began in June 1943, when some of the surveying still had to be done. The total labor force numbered seventeen hundred, including three hundred Han Chinese and Hui who lived in the vicinity of Xining, as well as army recruits and Tibetans. The Ngolog tribe refused to contribute any laborers.

The route of the road was totally uninhabited, so most of the laborers had to travel six to eight days to reach their assigned posts. They were each paid six bushels of grain. With the exception of the army recruits, most had no experience building roads, so it was up to the engineers to teach them.

The several hundred miles of marshy meadows on the Tibet Highway
contain geological formations of very dangerous decomposed vegetation,
which can act like quicksand

Since the work season on the plateau lasted only from May through October, we had to make the most of the short time. All of our supplies had to be shipped from Xining. We ultimately used twenty thousand tons of flour, sixty tons of dynamite, twenty thousand buckets, ten tons of picks and shovels, and a hundred tons of lime and other materials.

All of this had to be transported on Tibetan yaks. Each yak hauled about eighty pounds at most, and they traveled about fifteen miles per day. Since the yaks were essentially wild, they would sometimes bolt and lose all their cargo. On average, for every one hundred pounds carried by the yaks, forty reached the designated location.

Five hundred of us prepared to leave Xining, and we had a grand send-off. Few realized the hardships that lay ahead. We were to meet most of the laborers at assigned intervals before we entered the totally uninhabited areas.

We started with 355 saddle horses and 1,500 yaks to carry sixty tons of equipment, one hundred tents, and three radio transceivers. In fact, only about 150 horses survived the trip, so many of us ended up walking. Fortunately, Governor Ma took a personal interest in the road construction, and only his rigid leadership enabled us even to start the construction.

We crossed the Yellow River on a ferry built by our carpenters and eventually reached Payekala Mountain, the highest point on our road at an elevation of 17,275 feet. This was about the highest point in Qinghai Province, with the exception of Animaching Peak at an elevation of 23,405 feet. Considering that Mount Everest was 29,002 feet, this was indeed the roof of the world.

At an elevation of 16,500 feet, the air contains about half as much oxygen as exists at sea level. At 23,000 feet the oxygen is at the minimum amount needed for human survival. Our people could not adjust well to altitudes this extreme and usually developed altitude sickness, with headaches and difficulty breathing. We took great care with our horses, which were sturdy mountain ponies, and even fed them by hand with our own zanba, because we knew that if they died we had no replacements.

During the two years I worked on this road, I never met anyone other than our own crew. Except for an occasional radio communication with Lanzhou and Chongqing, we existed in a world apart. This was a world with no hostility, no personal property, and no civilization. We were committed only to the completion of the road and to survival in the same manner as the animals of the plateau.

− Chapter Fifteen
The Qinghai-Tibet Highway

In August 1944 we finally finished the road to Yushu. The 520 miles of highway were believed at the time to constitute the most difficult road project in the history of construction in China. At about the same time, the other crew completed the Xikang-Tibet Highway, and both roads now met at Xewu, a short distance north of Yushu.

When our report on the completion of the road reached Chongqing, the Ministry of Communication decided to send one bus and three of its Russian-made trucks to test the road. The inspection party included a high-ranking political official, a high-ranking engineer, a mechanical engineer, a photographer, a reporter, and others. We were told to meet the team at Xining and to drive with them to Xewu, where the inspection team from the Xikang-Tibet Highway would join us.

This inspection became one of the biggest events of my life. After more than two years of unbelievably hard work, our engineers could face only complete success. Governor Ma understood the importance of the moment, and he became as excited as we were. During the past two years, he had done everything humanly possible to help with the completion of the road. We welcomed the inspection party with a lavish dinner so we could all become acquainted. I told a brief story about the construction and the hardships we had faced during the past two years.

The next day we gave the inspection party a tour of the Taer Monastery, a replica of a great Potala monastery in Tibet that had a beautiful bronze-plated roof. The building was built during the Ming and Qing Dynasties. Perhaps its most distinctive attractions are the many statues that are made by mixing pigments with butter and molding the mixture to shape. Because of the high-altitude climate, these statues of colored butter can keep for years without deteriorating.

Then we began the journey, taking our caravan along the road for the inspection. One stretch of road that first day, about a quarter of a mile long, consisted of very deep fill. Because of our lack of rollers, the fill had never been well compacted. I had worried greatly that heavy trucks might not be able to get through without sinking.

I had decided the day before not to leave this matter to chance. Unknown to the inspection party, I ordered two hundred laborers to go to the area and soak this stretch of road with buckets of water. That night, the entire fill froze solid. The following day the caravan rolled over the fill without mishap.

When we came to the Sun and Moon Mountain, I worried that the trucks, with carburetors designed for lower altitudes, would have trouble in the thin mountain air. Luckily, the Russian-made trucks had plenty of power, and we made our way slowly up the seven percent grade.

We traveled 160 miles the first day and pitched our tents at Tahopa. Privately, I congratulated myself on a successful first day but did not dare become overconfident. If we crossed the Huanghe (Yellow River) safely, we would be past the major problems.

The Huanghe originates at Qinghai. The exact source is difficult to pinpoint, but it is probably in the area of Linghu and Chalinghu at an elevation of about 14,300 feet. Nokutsungliehchu lies to the west at an even higher elevation. This area could also be the source for both the Huanghe and the Yalungjiang. To determine the source on the ground would take another expedition similar to the search for the source of the Nile, although perhaps today we could use satellite photos.

By the time the Huanghe reaches Huangheyan, (bank of the Yellow River), its course becomes more stable. The water flows gently and clearly and has fed many small lakes located around Huangheyan. I recall that the great poet of the Tang Dynasty, Li Bai, wrote, "Water from the Yellow River comes from heaven, rushing toward the sea with no return."

We reached Huangheyan at noon, and the barge we had built for the vehicles was waiting for us. When we drove the bus onboard, we realized that the water had become so shallow that the hull of the barge nearly touched the riverbed. We had a great deal of trouble launching the barge into the river.

Since the river was too shallow for the barge, I figured that the big trucks could drive through it. That was a big mistake. Instead, the first truck bogged down and stuck fast in the middle of the river.

We were not immediately concerned, as anyone who has traveled in northwestern China was used to such setbacks. We got into the river and tried to push the truck free. As we worked, something unexpected happened. The temperature dropped twenty degrees within thirty minutes and continued to drop. When night fell, the temperature had reached ten below zero, and the river had become a sheet of ice that held the truck firmly encased within it. The wheels, the transmission, and everything else were frozen.

We gave up and pitched our tents for the night. However, I sent an urgent message to Governor Ma explaining the situation. I understood clearly that if we could not get the trucks across the river, our two years of hard work would have been for naught.

The inspection party began to grumble that evening. By daybreak Governor Ma, who had received our urgent message, sent fifty guards from Xining. They arrived on horses, carrying heavy ropes. Lashing the ropes to the vehicle, they tried their best to free the truck, but they could not move it an inch. As they worked, an elderly man in our party spoke up: "Your trouble is that you have offended the guardians of the river, the vultures."

In Tibet, the majority of the common people practice what they call celestial burial. The body of the dead person is covered in white cloth and is kept for three to five days before the funeral. The undertaker, a religious figure, dissects the corpse, cutting the flesh into small pieces. He crushes the bones and mixes them with zanba. Then he feeds the balls of zanba to the vultures and then feeds them the flesh. Vultures are sacred to the Tibetans, and they have become a protected species. Everyone is forbidden to shoot them.

"Now, your troubles began when one of your guards shot a vulture and offended the river god," the old man said to me. "Since you are the leader of this party, you must slaughter a sheep then kowtow to the god of the river to redeem yourselves."

Although I did not share these superstitious beliefs, I was desperate. Besides, making a good impression on the Tibetans seemed like a good idea. We had a sheep slaughtered, and I knelt down and kowtowed three times to ask for forgiveness. Then we tried to pull the truck out again.

For whatever reason, the "miracle" happened: with little effort, the mounted guards hauled the truck out of the riverbed to the opposite bank. The wheels of the truck did not turn, and many feet of surrounding ice came out of the river with the truck, all in one piece. With everyone cheering loudly, we knocked off the ice and started the engine. The rest of the caravan also crossed successfully.

I cannot comment on whether one should believe in the supernatural. In this case, my act of contrition apparently got the job done. Throughout history miracles have been recorded by all religions, and I think it would be a mistake to discount them all on purely scientific grounds. I believe miracles happen only in the minds of believers and that mere mortals have not even come close to understanding the mysteries of the universe.

After that perilous crossing our trucks had no more trouble. We finally reached the Payekala mountain range at an elevation of 17,275 feet, the highest point on our road. With an elevation of this magnitude, we claimed that the Qinghai-Tibet Highway was the highest in the world. I expected to see steep mountains reaching toward the clouds, similar to what we had seen on the Burma Road. Surprisingly, the world-renowned Payekala range was just a small hill in relation to the surrounding area. Building roads over the range had posed no real problems, yet it was certainly a challenge for those who had to climb the mountain range on foot.

Soon after we crossed the range we came to the Chuxe Monastery. This was the first man-made structure, with the first human beings outside our own party, we had seen since we had crossed the Sun and Moon Mountain. We did not stop long. I was eager to face the next challenge just ahead, crossing the Yangzie.

The Yangzie has at least two other names. In much of China it is also called the Changjiang, or Long River. Locally in Tibet, its upper reaches have the name Tangtianhe, meaning "Reaching Heaven River."

The source of the Yangzie was a subject of much debate for centuries. As early as 478 B.C., during the period of the Warring States, Mingshan in Sichuan was called its source. During the sixteenth century, scholars decided the Yangzie originated from the Jinsha River. In 1920 an expedition was sent to the Qinghai-Tibetan plateau to find the source but failed to do so.

In 1976 the exact source of the river was finally located by a scientific expedition sent by the Changjiang Valley planning office. They pinpointed the source at Geladandong Peak, 21,717 feet above sea level. The peak is the highest point in the Tanggulia Mountains, which border Qinghai and Tibet. At the highest altitudes the stream meanders along the eastern tilt of the plateau through wide, marshy meadows for around ninety miles before it crosses our highway at Tuotuo. From there, it continues east for thirty-six miles and is joined by the Damqu River, which carries a very large volume of water. From that point on, the little stream becomes the turbulent Tangtianhe.

By the time the Yangzie has traveled from its source at the lofty peak of Geladandong and emptied into the East China Sea, it has traveled 3,915 miles. The longest river in Asia, it is the third-longest river in the world after the Amazon and the Nile. At Tuotuo we found that unlike the shallow Huanghe, the Yangzie was about four hundred feet across and very deep, with swift currents.

Fortunately, the barge we had arranged for was waiting. This time we had no trouble getting the bus and the three trucks across the river. I have heard that a bridge now spans the river at this spot.

On November 13, 1944, our party reached Xewu, where we found the other inspection team from Xichang waiting. Our two-year ordeal had ended. The inspection team declared our road satisfactory. Finished at last, we proceeded to Yushu, only thirty miles to the south.

Today Yushu, which is located in the Tangtian River Basin, is the capital of the Yushu-Tibetan Autonomous Prefecture. In 1944 the town had a few mud houses and a couple of trees. The rare trees were considered more precious than jade, and so the town was named Yushu, meaning "Jade Trees." Today, new buildings and a forest of trees decorate the city, making it truly green.

After a brief rest in Yushu, we started our journey home.

In Chongqing, in 1945 I was voted the most outstanding engineer of the year by the Society of Chinese Engineers. This honor was usually reserved for bureaucrats with high positions in the government. I was delighted to receive it at age thirty-three.

Sometime later, I read a news release about the American inventor of the ballpoint pen, Mr. Reynolds, who claimed to have found a mountain in Qinghai that exceeded thirty thousand feet, as well as some active volcanoes. I was surprised, because I knew these were false claims. However, I was amazed to find that the Chinese public would believe anything just because it came from the United States.

During the same year, I was called for an interview by Generalissimo Jiang Jieshi. He asked me to recount my accomplishments on the Burma Road and on the Qinghai-Tibet Highway. Then he ordered a medal for me in recognition of my achievements. In the Guomindang at that time, these medals were usually political and were awarded only to high government officials. This occasion was the first time such a medal had been awarded to an engineer. I thanked the generalissimo and asked him if he remembered Chen Mu.

"Of course, he was my fellow revolutionist."

Then I told him Chen Mu was my father. In a moment that was more spontaneous than ceremonial, he rose from his seat and shook my hand. With that, my highway engineering days came to an end.

To survive in the grasslands on the Tibet Highway, one needed "a fast gun,
a swift horse, and a sheepskin coat"

The engineering camp at the Cluster Sea, the source of the Yellow River

Engineers taking a break in the snow in July

The Qinghai-Tibet Highway senior engineers (author in the middle)

Crossing the Yellow River on a barge (river width about 330 feet)

*Celebration of the completion of the Qinghai-Tibet and Xikang-Tibet
Highways at Xewu on November 13, 1944*

Chapter Sixteen

Civil War and Hong Kong

On September 2, 1945, V-J Day, China should have held a great celebration. The country had survived eight long, bitter years of war with Japan, and millions had died because of floods and starvation; millions more had died at the hands of the Imperial Japanese Army. Now the foreign invader had been defeated, but the war had not ended. I knew it, most Chinese knew it, and certainly the generalissimo knew it. Still, a small crowd gathered in front of the presidential office, hoping the generalissimo would appear and address them. He never came out.

From 1945 through 1949, I led a very quiet life. My son, Tyrone, and another daughter, Yvonne, were born. I taught at the University of Chongqing and later at Fu Tan University. At the same time, I had a full-time job in the municipal government of Shanghai.

In Shanghai I was in charge of a district that before the war had been designated an area for new development. During the war, the Japanese had demolished the modest amount of new construction they found. My efforts to revitalize the area were fruitless. The civil war between the Communists under Mao Zedong and the Guomindang under the generalissimo continued to rage, and the government had few resources for the reconstruction of Shanghai.

I fully realized my limitations as an engineer during that period. In fact, I worked only halfheartedly, doing just enough to meet the requirements of the bureaucracy. I spent my time writing a complete treatise on soil mechanics in Chinese, and I also prepared a book on a highway pavement study.

I learned more about the practical side of engineering in Shanghai than I had through all my education at the Universities of Michigan and Illinois. I finished my book, *Highway Pavement Study,* and sent the manuscript to the publisher just before the fall of Shanghai to the Communists.

In 1947 the United States asked the Ministry of Communication to send twenty top highway engineers to the United States for further education and training. The ministry was happy to comply and placed my name at the top of the list. After several reviews of my record and some interviews, the ministry decided I should lead the group. An engineer from the United States came to Shanghai for the final round of interviews, and he approved the names on the list. I was reluctant to leave my wife and children, but this group was important to the future of engineering in China.

A week before our scheduled departure, the head of the Ministry of Communication decided to interview all twenty engineers personally. Through a peculiar working of fate, he was no stranger to me. He was General Yu, who had fled from the advancing Japanese on the Burma Road during the night without taking time to dress.

In the interview we all sat around a large conference table, with General Yu and two vice ministers at the head. The general looked at the list and called us one by one. When my name was called, I stood up, and he looked at me intensely for a moment. "You cannot go; your complexion is not good," he said. Everyone was surprised.

The vice ministers had known me well for many years, and one pleaded my case. "Mr. Chen has served the country extremely well. He has gone though many hardships, especially on the Burma Road and the Tibet Highway. We see no reason why he cannot go."

The general banged his fist on the table. "My decision is final. The man's complexion is bad, and I know what I am doing. I am a physiognomist."

At that point I lost my patience and barked, "Your honor, sir, I don't care if I go or not. I've been in the United States, and I don't care to go a second time."

Startled, he didn't answer right away.

"Sir, do you remember that time when you fled from the Japanese without your boots?"

That was too much for him. He motioned to his guard and said, "Arrest this man, and put him in jail." That was the end of my interview.

As it happened, I knew people in high places who were better placed than the general. As a result, I was released immediately. Later I learned that the general had become demented from the pressure of his job and was living a quiet life in Taiwan.

Needless to say, I did not go with the group of engineers to the United States. That was fine with me. My wife was delighted with the outcome of the episode.

I did go to Taiwan for a short time to take over the highway network left behind by the Japanese. I expected to see some good engineering and an efficient operation. Instead, I found single-lane roads of substandard construction. In later years, after the Guomindang government had moved to Taiwan, the government built superhighways that completely crisscrossed the island. These proved to be difficult jobs that the Japanese had never dreamed of accomplishing.

During my short stay in Taiwan, the provincial government urged me to stay and take over the Highway Administration. I politely declined. At this time I needed to open a new chapter in my life.

In May 1949, the Communists were approaching Shanghai, one of the last bastions of the Guomindang in the civil war. I was asked to address the entire Public Works Department about the progress of the defensive work around Shanghai. Instead, I blew my top.

Standing before the audience, I condemned the government for destroying the castles of idealism for which my father had given his life. I accused the bureaucrats of corruption at the highest and lowest levels. Angrily, I railed against the suppression of the poor for the benefit of the few. Finally, in tears, I finished with the blunt truth: "So be it. We will see the map change color very soon." The entire audience was silent, including my boss.

On May 25, 1949, the Communist forces arrived outside Shanghai. Several months earlier I had sent my family to stay with my mother in Fuzhou. I asked myself whether I should go to Taiwan with the generalissimo and his government or stay in Shanghai, possibly to begin a new chapter in the development of my native country, with new concepts and new goals in the government that was about to take power.

My mind was made up as a result of new information. I found out that my best friend and most faithful employee for fifteen years was an agent of the Communists. He had been reporting all of my activities to the Central Committee, and, of course, I had been working for the Guomindang ever since I had returned from the United States. I had had enough. I took the last plane out of Shanghai, intending to go to Taiwan.

On the plane I decided I would no longer have anything to do with the old regime. People like General Yu, for whom I had no respect, would continue to run Guomindang politics, and the inevitable outcome was a Communist victory in China. Was Hong Kong the place for me? I doubted it, but going to Hong Kong provided a first step in changing my life.

I left the plane at Fuzhou, where my family waited. The Communists had not yet reached it, and we still had time to leave.

We packed our belongings, and I hired a sampan for the trip from Fuzhou to Hong Kong. During the trip, I decided this was a foolish move. My wife and three children were with me, the oldest child now five years old. All of our luggage was piled on the sampan, and I knew nothing about the boat or the owner. Ugly stories circulated about sampan owners who killed all their passengers or forced them into the sea and ransacked their possessions. These stories weighed heavily on my mind during the two-day trip, so much so that I dared not close my eyes to sleep. I didn't tell my wife of my concerns, because she would have become upset, and there was nothing we could do now.

Fortunately, our host was honest and took us safely on our way. I felt a great sense of relief when I saw the skyline of Hong Kong. We docked at a small fishing wharf at North Point, and I paid the sampan owner generously. Now I had to decide what to do next.

Hong Kong had become a madhouse of activity; the population had doubled after the end of the Sino-Japanese War, and refugees from the Chinese civil war, like ourselves, were pouring in. All the hotels were jammed, and the rates were outrageous. I had very little money left, and everything seemed to be measured in gold.

I remembered that one of my fraternity brothers at the University of Michigan now had a small manufacturing plant in Hong Kong. So I called T. F. Poon on the phone and explained that I had just arrived with my family and that we were desperate for help.

"Come on over," he said without hesitation. "We'll fix you up."

I was pleasantly surprised; this was like an outstretched hand to a drowning man. Poon had a small shop in the front of his establishment and an empty warehouse in back. The warehouse had no furnishings, but we stacked up planks for a bed. All five of us slept on the crude pile of planks. It was not comfortable, but it was safe and was better than sleeping on the street.

The next morning I began my hunt for a job. A friend who was an architect had recently obtained a job designing a twenty-four-story building in Hong Kong, so I went to see him. To my surprise, he asked me to design the entire structure. As a highway engineer, I had not touched structural design since my school days. I was afraid I might not be able to do the job, but of course I was desperate. Still, I could hardly believe my ears when I heard myself accepting the assignment.

For more than a month I dug through my old textbooks and pored over everything pertinent I could lay my hands on. I had no desk, so I worked on a stack of cartons day and night. Finally, I finished the design, and my boss accepted the results.

I also had a complete surprise concerning the book manuscript I had sent to a publisher in Shanghai. The publisher had printed my book and had allotted a substantial amount of money for me. I wrote to the publisher and asked that the money be sent to my mother in Fuzhou.

My wife managed to get a job as a mathematics teacher at St. Joseph's Girls' College. This was a giant step for us, because it convinced us that we could make a living in the British colony among two million refugees.

We were beginning new lives. I wanted to forget the Japanese invasion and all of the bureaucracy and corruption that had ruined China. The time had come to forget the past and build a future.

Quite a new life it was, too. We sent our two daughters to a British school, and our son attended a Chinese school. Soon we rented a two-room apartment — it was crowded, but it was home.

One day I saw a Public Works Department ad for engineers in the paper. I applied, but I expected they would have hundreds of applicants, so I was surprised when I was selected. The pay was good, and I thought I would have a regular place to hang my hat at last.

I was assigned to the Road and Bridge Department, in which all of the senior engineers were British. At the time, the Hong Kong government was building roads into the so-called New Territories north of the original colony that had been ceded to the British after the Opium War. They supposedly needed new road engineers for the expanded system.

All of the survey work came under the purview of the Road and Bridge Department. My first job was to join a surveying crew, and I assumed that with my years of experience in survey and construction, my new assignment would utilize my skills. I couldn't have been more wrong.

The engineer I was to accompany was a new graduate from a university in Britain. I was assigned to hold an umbrella over his head, since it was June and this British engineer did not want to get a suntan from the intense tropical sunlight in Hong Kong.

At first I didn't really mind this aspect of the job. After all, I had lost my homeland and was living in a British colony. I was willing to work my way up in the department, and I could tolerate a certain amount of humiliation on the way.

I suspected my young boss had not had a single day of field experience before this project. One day we faced a small hill, and he became bewildered. He contemplated cutting right across it, and I could hold my tongue no longer. Politely, I suggested that we could easily go around the hill by moving the roadway to the right.

He looked at me and said, "What do you know about highway location? This is a mountain road and a difficult job, a job for a top engineer like myself. You hold the umbrella and let me worry about road location."

I held my tongue, but I went home ruminating about my position. I asked myself, why should I take this? Is it worth all the grief? Finally, I decided I had received too great a slap in the face, so I sent in my resignation the next day.

The commissioner of the Public Works Department called me as soon as he received my resignation. He was, as the British called them, an "old China hand" who had been in Hong Kong for many years. I believed he understood the Chinese better than most British people understood them.

"I am well aware of your past accomplishments in China," he said. "But mind you, you are living in Hong Kong now under British rule. We have one million people looking for jobs just to survive. My advice is to forget about your past. Take whatever we give you and be grateful."

He was right. I should have swallowed my pride and taken whatever bone they decided to throw my way. Too many people in Hong Kong were desperate.

"I am transferring you to the port work office," the commissioner continued. "There you will have a chance to establish a material-testing laboratory for Hong Kong. Now go on, and no more nonsense."

So I accepted the new job, which at least allowed me to avoid that young engineer. Within two years I began to make my mark in the department. I established the first material-testing laboratory in Hong Kong and became the highest-paid Chinese engineer there. My wife was promoted to a position as a science teacher at St. Joseph's, where she was well liked and highly respected. She orchestrated the first science fair for the school and received the top salary among Chinese teachers there.

I received a letter from my former boss, Mr. Zhao, asking me to return to China to take charge of an important post. At about the same time, the Guomindang sent an agent from Taiwan, where the generalissimo's government had fled from the Communists, to see me. He gave me many reasons I should remain loyal to the Guomindang and return to Taiwan.

By that time I had already decided that I had no intention of returning to either China or Taiwan. Also, as far as I was concerned, the Guomindang administration, which was now in Taiwan, had deserted me. I could not really blame them, since they could barely save themselves under the circumstances. However, I had learned to survive rather well in Hong Kong, with a small but comfortable home and good schooling for my children.

I had successfully begun my second life in Hong Kong. Sometimes I still had dreams about my days working on the Burma Road and the Tibet Highway. Yet those days seemed remote and foggy, almost as though they belonged to someone else's life. At age thirty-nine, I still had half of my life before me, and I wanted to go forward, not backward.

I could predict my future life in Hong Kong fairly well, certainly better than I might have anticipated. A Chinese bureaucrat in Hong Kong could only go so far. I was in charge of the material-testing laboratory, and since the British Civil Service had clear guidelines for such service, I knew I had reached the highest level I could hope to achieve. If I chose to stay in Hong Kong, I would have to be content with my present pay and with always serving the whims of a British colonial officer. My job would be soft and steady, without challenges. My children would be raised as colonial offspring. Ultimately, their lives would be the same as their father's.

Would I be content to spend the rest of my life in this fashion? I woke up frequently in the middle of the night, wrestling with the question. I discussed the matter with my wife, as my children were not old enough to have informed opinions.

All of a sudden I made up my mind. I wanted a new adventure and that meant going to the United States, the land of opportunity. The United States would not only offer me new opportunities but would also offer wider horizons to my children when they grew up.

When I inquired, I found that the United States was accepting refugees from China on a quota basis. My wife agreed that we should try to go. I knew we would need some luck, since the American Consulate in Hong Kong was besieged with applicants.

The red tape began. I had to fill out a weighty tome of inquiries beginning with my childhood, reporting everything I had done since and discussing everyone with whom I was acquainted. Then we came to the most important part, the interviews. I was escorted to a small cubicle, much like those in the movies used for interrogation.

The interviewer asked all kinds of questions, many of which were totally unrelated to my immigration into the United States. After the first day I erroneously thought it was all over. I had no idea that this was just the beginning of thirty similar interviews yet to come. I began to wonder whether all of the hundreds of other applicants went through the same red tape I had. From what I could see, some of the applicants could barely speak English, and I wondered how they managed in the interviews.

This process dragged on for many months. Then came the prize question: "Do you know any Russians? Now, do not answer too quickly. Do you know any Russians, or are you even remotely acquainted with any?"

"Well, in my school days at the University of Michigan, one of my professors was a Russian."

"Aha," the interrogator said. "What was his name?"

"His name, as I recall, was Dr. Timoshenko."

"How old was he, and was he a citizen?"

"I really don't know. All I knew was that he was a world-renowned theoretician of engineering mechanics."

"Well, I can't let this probable Communist go. I will put a tracer on him right away."

At this point I lost all patience. This was insane. I had swallowed far too much nonsense in the past months.

"Look here, I have a good job in Hong Kong, and I don't have to emigrate to the States. The reason I applied was that I thought I could do something good in the United States, in return for what the United States has done for me in past years. If you have any doubts about my integrity, why don't we just forget everything? I'll withdraw my application."

He looked at me with an open mouth and did not say a word.

I got up to leave. "However, I think my past months of dealing with you should be recognized, so I will be happy to write up the entire business and send my report to *Life* magazine for publication."

I left his office feeling a sense of relief. I felt stupid for taking this treatment for so long. Then, the next day I got a call from the American Consulate, informing me that my application had been approved. All the papers were ready for me to pick up.

For years, I had harbored the illusion that in the United States everything was governed by law. From my experience during the years of my education in the United States, I believed government employees were uniformly law-abiding and that they did everything by the rules. I thought corruption was only a feature of our ancient bureaucracies in East Asia. Now I knew I was wrong; my threat to publicize the ridiculous interview process had carried more force than I had expected. Suddenly, I understood how hundreds of Chinese had acquired their immigration status so easily without the mountain of red tape I had been forced to climb. At least I had received my papers without spending a single cent to grease the wheels.

I had no intention of living in the United States without a job. I started to apply for jobs while I was still in Hong Kong. I wrote letter after letter to every conceivable location and never received even the courtesy of a reply.

During all the years I had spent in China and Hong Kong, I had remained a member of the American Society of Civil Engineers. As a last resort I placed an ad in the positions-wanted section of the group's newsletter. I was delighted

*Author (left) founded the Public Works Department testing
laboratory in Hong Kong, 1949*

when James Sherard of Woodward, Clyde and Associates, a company of
consulting engineers, offered me a position as a project engineer at a
handsome salary. I answered the letter at once and told him the date I would
disembark in San Francisco.

On January 8, 1957, I left Hong Kong on the SS *President Cleveland.* Thirty
minutes before the plank was lifted, a man rushed on deck and found me.
"Mr. Chen," he said, "you probably don't remember me, but years ago you
did some consulting work for me, and I never paid you." He handed me a
brand-new Leica camera and said he hoped it would compensate for what he
owed me.

Before I could thank him, he had hurried off the ship. This was quite a
contrast to the situation later in my career, when I discovered that I sometimes
had to resort to lawyers and lawsuits to collect what others owed me.

As I returned to the United States, I remembered the children dancing and
singing "Chin Chin Chinaman" at me when I had first arrived for college and
recalled the sign in the window of a restaurant saying "We do not serve

Chinamen." I remembered my humiliation when I could not find a rooming house on my college campus in Michigan. At the time, I had felt I had no choice but to tolerate such treatment as the price of my education. Now, however, many years had passed, and I was no longer a young college student. This time, I swore I would prove myself.

– *Chapter Seventeen*
Life as a Consultant

I arrived in San Francisco on January 26, 1957, at 10:00 A.M., returning to the land I had left in 1936. After twenty-one years I returned with little money but abundant ambition. I intended to put the past behind me and start a completely new life, and I was firmly convinced that this land would treat me right.

I was forty-five years old, an old man by Chinese standards. By American standards I had passed the prime of my life. However, I still felt young enough to face the new challenges before me.

I was grateful to Dr. Sherard, who had hired me without an interview. Without his endorsement I could not have left Hong Kong. Unfortunately, Dr. Sherard was not stationed in Denver, Colorado, the city to which I was to report. At that time Denver was a small western city, a so-called cow town. It was growing rapidly, however, and expected to continue expanding. Best of all, the people lived up to one of the better legends of the American West: they were friendly.

Woodward-Clyde, a geotechnical consulting firm, had offices in many major U.S. cities. When I arrived, the engineering staff in Denver consisted of only four people: an Israeli who had graduated from the University of Colorado, an ex-military man from West Point, a good-hearted young man who talked incessantly, and a former government employee. The last of these was my boss. He was a stern taskmaster who expected his engineers to handle a project all the way through, from the basic fieldwork to laboratory testing, writing the report, and billing the client. In other words, no assistance, including manual labor, could be expected. As a result, I learned a great deal about the workings of private enterprise and how to handle business.

With this experience I came to see that private enterprise really is the backbone of U.S. business, as the cliché says. The only way to make money is to be competitive, and the only way to be competitive is through efficiency.

Working for Woodward-Clyde helped my subsequent business career in incalculable ways.

In fact, the first lesson I learned as a businessman in the United States was that the primary objective in business was to make money. My main objective as an engineer in China had been to get the job done within the specified time. Profit had never been a primary consideration, as long as we had the money to finish the particular road. In my new job, each engineer had to produce at least three times his salary to justify his continued presence in the company.

I was to be naturalized as a U.S. citizen five years after I had first been granted permission to enter the United States from Hong Kong. Even now, I vividly remember the day before the naturalization ceremony, when I was in Julesburg, Colorado, working on a project. Julesburg is a small town near the state border, about two hundred miles east of Denver. I rose at the crack of dawn and drove back to Denver for the ceremony. My wife was afraid I would not get there, but I reached City Hall just before my name was called.

Meanwhile, the Communist regime in China had consolidated its control on the mainland. The Soviet Union had supported the Communist Chinese for many years and was the first country to have diplomatic relations with China. During this time, the United States did not recognize the new government in China diplomatically.

In China, everything from Russia was now considered good and was to be highly valued. Chinese students learned to speak and write Russian. Factories in China employed Russian consultants. Slogans of Marx and Lenin were put up on billboards and street corners.

As a result, many Americans became convinced that the Chinese-Soviet bloc would ultimately rule all of Asia. U.S. Secretary of State John Foster Dulles expressed this idea in his domino theory. The central idea was that if communism wasn't stopped in each country, all others nearby would fall like dominoes and be integrated into the Communist sphere of influence.

Dulles had brought the United States into the Korean War because of his domino theory. Technically, the Korean War was a conflict between Communist and non-Communist forces that lasted from 1950 to 1953. China entered the war in October 1950. The conflict eventually evolved into one between the United States and China. Both countries and, of course, the two Koreas suffered heavy casualties, but the Korean War, following the successful Communist revolution, raised the morale of the Chinese.

I was invited to speak to the Mining Club, before an audience of several hundred people. In my talk I boldly predicted that China and the Soviet Union would split sooner or later. This was an unheard-of opinion, and the *Rocky Mountain News* reported my views as "Chinese engineer predicts the breaking of ties between Russia and China." My boss told me I should stick to engineering and leave politics to those who knew what they were talking about.

Only a few years later, Mao Zedong did split from the Soviet Union, to the surprise of everyone, including the "old China hands." The split was reported as political in nature. During the split, the Soviet Union shipped millions of tons of valuable mineral resources from China to the Soviet Union. With the loss of resources and Russian consultants, engineering projects in China were left unfinished, with blueprints taken away or destroyed. The actions taken by Mao were draconian, but this split from the Soviet Union ultimately proved to be good for China.

The attitudes of the Chinese toward the Caucasian race changed throughout the period from the Opium War to the cold war. During the early Qing Dynasty, the whites were considered "white devils," an inferior race. When the Qing Dynasty was soundly defeated by the British in the Opium War, this view changed drastically, and the whites were thought to be superior. Everything they touched or did, by definition, had to be good. The Chinese could do nothing right except to kowtow to the whites. I was raised during this period. We were taught from early childhood to follow Western cultural ways, to seek Western knowledge, and to abandon traditional Chinese ideas. Students of my generation were sent to European and American schools for their education. Anyone who completed a Western education was considered to be an accomplished intellectual, and he or she would hold important positions in the government; my own quick welcome by the Chinese Highway Administration in the 1930s reflected this belief.

The Communist revolution had swept away all of the foreign concessions such as the one in Shanghai, where I had lived as a child, excepting only the colonies of Macau and Hong Kong. When the Chinese fought the West to a standstill in Korea, the Chinese finally realized the whites were not as formidable as they had been led to believe. Everyone, regardless of their political views, realized that China's century of humiliation had finally come to an end.

I had learned a great deal about U.S. business practices and about how private enterprise could accomplish many tasks better and more efficiently than a top-heavy bureaucracy such as the Guomindang in Chongqing. Also,

I had reaffirmed my faith in the fact that hard work could bear fruit in the United States. My years as a consultant with Woodward-Clyde had shown me many things I would treasure for the rest of my life.

After two years with Woodward-Clyde, my boss called me into his office one day. He said, "You know, at one time I almost fired you, but since then I think you've gradually gotten hold of how to be a consultant. I'm going to put you in charge of engineering."

Naturally, I was honored. Actually, I thought I already had an excellent position with the company and was fairly happy with what I had been doing. This promotion told me I had made a good transition to engineering in Denver.

During these years I learned some fundamental differences between Chinese and Americans. The Chinese tend to be more subtle in expressing themselves; they would never tell you straight out, for instance, that you had nearly been fired but would approach the subject in a roundabout way. Americans seem to get mad easily and then forget the entire matter. The Chinese tend to hide their feelings deep inside and never forget. Because of this, I think, it was rather easy for Americans as a people to put aside the attack on Pearl Harbor and become friends with the Japanese. I, for one, could never forget what the Japanese had done to us in China. When I closed my eyes, I could still see the Japanese soldier pointing the submachine gun at me. I seriously doubt that China and Japan can ever become truly friendly nations.

Our family life also settled into a comfortable routine. I bought a house with a carport and a second-story bedroom, under heavy sales pressure from a real estate agent. My new neighbor lost no time in selling us a vacuum cleaner, and his son offered to mow my yard for a modest fee. I had yet to learn how to say no in this world of highly skilled salesmanship.

Edna found a job teaching mathematics at the University of Denver graduate school. She was the first person without a Ph.D. to hold such an appointment there. My son, Tyrone, and my daughters, Dorothy and Yvonne, enrolled in the public school system, so it appeared that my life was well mapped out for the future as a middle-class U.S. citizen.

Even so, I was restless. In an attempt to figure out why, I asked myself what I had accomplished thus far in my life. I had failed as a son; my mother had died at age ninety while I was in Denver, and according to Chinese tradition the worst breach of filial loyalty a son can commit is to be absent during the death of a parent. In China I had failed as a citizen, as measured by the Nationalists' expectations, and the Communists considered me a Nationalist. I had failed as a businessman by U.S. standards, having worked only for others. Further, I could not find comfort in religion, because I stubbornly refused to believe in God within the confines of a formal religion. I could not immerse myself completely in life in the United States because of my deeply rooted Chinese upbringing.

By now I had realized that I had no particular talent other than engineering. I had become imprisoned by my own ideals, which seemed to have no place in the mad rush of pragmatic U.S. society. Where should I go from here? Was this settled routine to be the end of my journey? I wished I had the answers to these questions, but I did not.

An event that was out of my control helped me find the answers. My boss elected to hire a new employee, a former Army Corps of Engineers man, as chief engineer. The boss relieved me of all my duties as chief engineer and told me to work under the new employee. Was I not doing my job? Or did the clients not like to see a Chinese handling their projects? I never did find out.

Anyway, it was quite a blow to my pride, and I decided to quit the firm. My boss was upset and promised to increase my salary and benefits. This offer did not persuade me to stay, because I had made up my mind to try a new venture, one that to my knowledge had never been attempted by a Chinese engineer in the United States. I had decided to start my own business.

– *Chapter Eighteen*
Chen and Associates

In February 1961 I started my own business, Chen and Associates. I was forty-nine years old. It was the adventure for which I had become restless.

When I told my boss at Woodward-Clyde my intentions, he sat back and laughed. "Do you know how Dick Woodward got started making his money? Do you know how much money he had to invest to start his business? Do you know the kind of social prestige he had in California?"

I admitted I knew nothing at all about it. It was fortunate I didn't; otherwise, I might not have had the nerve to found Chen and Associates. Woodward had possessed many advantages I lacked.

Among the things I didn't know was that I couldn't afford an office, much less a secretary. I couldn't afford to buy testing equipment, and I was handicapped by the fact that I didn't want to steal clients from my former employer. In fact, I knew no one I could count as a potential client, and I could not afford to be without an income for even a couple of months.

We held a family meeting to discuss the matter. All of my family members were against this venture. I had only one thing going for me — the fact that I had not yet failed in any of my ventures. I decided to do it.

The first order of business was to locate an office. Since I could not afford to rent office space, I decided to convert my carport into an office. With my scant knowledge of building construction and with the help of my neighbor, we started the remodeling. I painted the company logo myself.

We picked up the bare essentials of office furniture at a Ford Company rummage sale. I could not afford a copying machine. To simulate one, I used a pickle jar filled with ammonia and inserted the typed sheets one at a time.

For my typing I relied on my daughter Dorothy, who was studying at the University of Colorado at Boulder. She came in on the weekends and did all the typing. I made calls to potential clients and did fieldwork during office hours. When my wife was not teaching at the University of Denver, she handled my phone calls. Thus, my first year of business was completely a family affair. I enjoyed the work, and the few jobs I obtained were enough to support the family.

As it turned out, the least of my problems was the lack of testing equipment. I was able to get by with low-cost substitutes for much of the expensive machinery. After all, during my days in Xian I had created makeshift equipment with whatever I found at hand. Now, the most expensive item was an oven for drying. I simply used my kitchen stove. Compared with those days in Xian, this was pure luxury — I had running water and electricity.

For a business to grow, capital can be acquired by issuing stock. I decided to issue five thousand shares at a dollar per share. When I rounded up all the money I had, I remained two thousand dollars short. I asked my good friend the driller whether he was interested in buying any shares. "I already gave to charity," he said.

I got similar answers from everyone I knew. Finally, I gave up and sold my stamp collection to make up the shortage. I knew my collection was worth much more than I received for it, but I had no choice. Now I had my required shares, and I received my formal registration with the state of Colorado as a business entity. I would never have dreamed that twenty-seven years later my stock would sell for $1,000 a share.

To get my business firmly established I worked long hours, seven days a week. Although I had little success at first, I believed for the first time that I was beginning to understand U.S. business life. From the perspective of a foreign-born American like me, the average American man lived on wheels, ate at fast-food places, worked around the clock, was ruled by the fair sex, and was tyrannized by his children.

Money seemed to be the central motif in American life. As youngsters, Americans had to earn "pocket money" to appreciate its value. Enrolling in college was considered important, because a college degree meant one would make more money. Hobbies like collecting stamps or coins were no longer a matter of relaxation but were investments. People changed jobs for just a small increase in wages, and before they died they selected the mortuary that gave them the best deal.

My Chinese immigrant friends told me I was working on a lost cause. They insisted that white Americans harbored deep prejudices against Asians and would send me no work. I refused to believe them, because I believed Americans were a fair people.

If I could do a better job than my competitors and deliver the work on time, I could always beat the competition. The major problem, as I saw it, was to become known in the business. To that end, I made this outline:

* Obtain recommendations by word of mouth
* Obtain recommendations through engineering societies
* Publish in engineering journals
* Avoid bad-mouthing fellow engineers
* Respect the opinions of others
* Give out advice freely even when no pay is in sight
* Quote reasonable fees
* Always deliver the product at the specified time

I firmly believed that if I could meet these criteria, I had a good chance for success. Establishing name recognition in the business took time, however. As a result, my first year in business was the worst I experienced. After that first year people began to murmur, "There's a Chinaman in town who can do a reasonable job. Why don't you give him a try?"

One of my main competitors had been a fellow worker at Woodward-Clyde several years before. He had started his own company with sound financial backing, and he possessed a keen business sense. One day he called me and said, "Chen, shape up. You know I could ruin your business in one week if you don't shape up."

By shaping up he meant price fixing — that I should raise my rates to meet his. I did not argue with him at the time and politely ended the conversation. Yet years later he asked me whether I wanted to buy his business. I had learned from the old Chinese saying that the "true man will tolerate the immediate insult." By that time he had become one of the richest men in Denver; however, he did not accrue his fortune through engineering but through dealings in stocks and real estate.

During my first two years of business, I pretty much did everything myself. I dug trenches with a pick and a shovel. When necessary, I entered crawl spaces on my hands and knees, and I examined sewer outlets and drainpipes.

Once, when I had finished drilling a large-diameter caisson hole for a downtown office building, no one would go into the hole to inspect the work. At that time there were no EPA or OSHA regulations about safety, so no straps or harnesses were required for inspectors. I rode the driller's Kelly bar down the sixty-foot bore to the bottom of the hole. The view from the bottom was scary, as I looked up to see the sky the size of a dime and thought about the tons of earth all around me. However, the hole was fine, and I survived the trip without incident.

Both the driller and the client were so impressed that the next day they sent me an affidavit stating that I was so devoted to my work that I had entered the hole to inspect conditions when no one else would. The document was duly signed and sealed by an attorney. After that, this client wouldn't deal with any engineer except me.

I find the present U.S. attitude toward racial labels perplexing. In fact, I don't really mind if someone calls me a Chinaman. No person can gain respect through legislation. Perhaps legislation can exercise control over what people say in public, but it cannot change what they think.

I began to realize that in Denver, people did not discriminate against me because of my race. They seemed to enjoy doing business with me. I slowly began to forget that I was of a different race and began to consider myself an American and an equal.

In December 1962, the state of Colorado issued the certificate of incorporation for Chen and Associates, complete with by-laws and a seal. Also at that time I joined the Rocky Mountain Toastmasters Club. The people in the club helped me tremendously in establishing my business. For instance, Glen Taylor prepared all my legal work without charge; Don Wagner gave me the best deal available on insurance. Bill Bowen had given me my first job, and Al Anderson referred me to my first industrial client. I found that these Americans were willing to lend a helping hand to their fellow businessmen, especially to someone just starting out.

As my sales increased, I decided I could afford an office. I was glad to get away from my homemade office in my carport. Also, for the first time since I had started my business, I was able to hire employees to share the fieldwork. My first choices were a man from Thailand and a man from Iraq, Sid Siddeek. They proved to be excellent choices. They both spoke English with a heavy accent, yet I found my customers trusted them more than they trusted some of the smooth-talking local engineers who were better at selling than they were at engineering.

The Thai engineer quit after five years and is now operating a very successful Thai restaurant. Siddeek, who was in charge of a branch office for several years, is now teaching Iraqi at a California college. He has no love for the present Iraqi regime and has remained in the United States.

When Eastman Kodak decided to move part of its Rochester, New York, plant to Windsor, Colorado, the company selected geotechnical firms and came up with three names: Dames and Moore, Woodward-Clyde, and Chen and Associates. The first two were internationally known companies, and I was not surprised that they were in the running. For a start-up company such as my own to be listed with them was a complete surprise, and I had no idea how my name got on the list.

I thought there might be some mistake, but I asked my wife, "What have we got to lose? Kodak is paying for everyone's expenses to travel to the interview in Rochester. We might as well go to Niagara Falls for a short vacation."

So we went. I assumed that the interview would be about an hour long, but it started at nine in the morning and, with a break for lunch, it continued until four in the afternoon. Much to my surprise, eight people interviewed me, three of whom were college professors. They asked me questions about basic soil mechanics and about my general experience as an engineer. When the interview ended I felt I had handled the matter well, but they gave no indication of their opinion. We had a good time in Buffalo and then returned to Denver.

I expected Kodak would select one of the internationally known companies and gave the matter no more thought. Then a letter came to announce that Chen and Associates had been chosen to do the project. I could not believe my good fortune.

Immediately, I began to prepare for the work, so we could do a good job for Kodak. Many years later I learned a little about the selection process. Kodak at the time had many Chinese employees and had been very happy with them. The company assumed I would do a good job.

Receiving the Kodak contract was a turning point for Chen and Associates. From that time on, whenever clients asked about my experience, I told them Kodak had used my services. This made an excellent impression. I landed one big project after another, and the name Chen and Associates became well-known in the circle of Colorado engineers. I do not know how much I owed to mere good luck and how much to hard work.

In the Rocky Mountain region, property owners and construction companies had long been bothered by heaving foundations, which caused ugly cracks and occasionally resulted in structural failure. It was not until 1950 that the matter had been researched by geotechnical engineers. Locally, the soil was referred to as bentonite. Technically, it is an expansive soil.

I had always been interested in such soils. I devoted a lot of time to studying expansive soils, particularly those beneath high-priced homes and major structures that had suffered great damage. Many other engineers were researching the same problem, and we exchanged information at professional meetings.

The problem of expansive soils had been increasing steadily, not only in Colorado but also in Wyoming, South Dakota, Texas, Utah, and many other

Family house carport converted to author's first company office in 1961

states. As the demand for new homes had increased, new real estate had been chosen for development. Consequently, land rich in expansive soils was increasingly used by unscrupulous or uninformed contractors.

In Colorado, engineers recommended the use of drilled piers for foundations, rather than the traditional spread footings. Still problems persisted. The legal profession viewed these problems as potential lawsuits. Lawyers often approached the owners of cracked houses built on expansive soils and offered to work on a contingency basis, where no fee would be charged unless the lawyer won the case. Owners liked this approach, because it cost them nothing and gave them the possibility of a reward. Suits were filed against contractors, architects, engineers, subcontractors, and even real estate companies. In the case of a $50,000 house, the damage claims could sometimes mount up to $1 million.

Those in the consulting professions were aware of the soil problem, but they could not provide an answer based on the knowledge available at the

Chen and Associates Oriental-style office building, 1974

time. They spent a great deal of money and time, in many cases paying over one hundred times the fee they had charged, just to settle the lawsuits. Any engineering company with substantial assets would not even touch projects involving expansive soils.

I devoted considerable time and money over a period of many years in an attempt to resolve the problem. Gradually, I acquired a reputation as a leading expert on expansive soils. In 1975 I published a book entitled *Foundations of Expansive Soils*. At the time, this was the only publication in the world that dealt with the subject, and it was widely circulated. As a result of the book, I received many letters from all over the world asking for advice. Ironically, lawyers from all over the United States wanted me to testify as an expert witness. In most of the cases I declined; I did not want to play the lawyer's game and testify against my fellow engineers.

After I had been in business for four years, Chen and Associates had grown from a one-man company to a sizable organization; from the office in the carport, there was now a respectable office in the main business area of

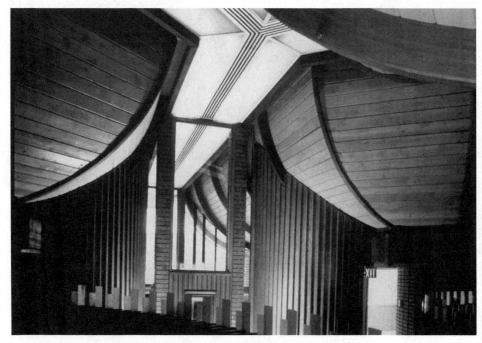

Interior of Chen and Associates office building

Denver. Now I truly felt I had become a part of the United States and the U.S. business circle. I had all but forgotten that I am Chinese and an immigrant. I joined and served in various organizations, and I spoke to engineering societies in Denver, as well as to service organizations such as the Lions, Sertoma, Optimists, and Kiwanis clubs.

Frankly, I have never believed any special consideration should be given to any minority. I believe that as Americans, we should all enjoy the same opportunities and share the same responsibilities. As a nation, the United States was built by immigrants. We should not now separate everyone by race, ethnicity, religion, and all of the other special-interest groups we have at present. The claim is that we identify and isolate these various groups initially to help them to compete, but eventually they become nothing more than self-interested groups, making exaggerated and disproportionate claims on our public resources.

At one time insurance companies refused to insure geotechnical engineers because of their high risk. To fight against this practice, in 1968 we organized ourselves and founded the Association of Soil and Foundation Engineering

(ASFE), with our own insurance company, Terra Insurance Ltd. Terra is an offshore association. As a captive insurance company, it provides standard practice policies for most types of engineering companies, provided the firm is a member of ASFE and meets its underwriting criteria. I was one of the founders of the ASFE. Today, insurance companies that once scorned the idea of insuring us are fighting to get into what they now see as a lucrative market.

In 1964 I decided to build my own office building. I saw it as an opportunity of a lifetime, to erect a landmark design in Denver. For this building I engaged Nixon, Brown, and Bowen as my architects.

Bill Bowen had given me my first job. Not only was he familiar with my business, having watched it grow through the years, but he had also been to East Asia and was familiar with Asian architecture. I wondered at the groundbreaking ceremony whether my business would continue to grow and whether I would be able to make the monthly payment I owed the bank.

I made the right decision. My building, which was designed in Asian architectural style, attracted a great deal of attention, and my business undoubtedly benefited from the publicity and prestige. By this time I had opened offices in Casper, Wyoming; Colorado Springs, Colorado; and Salt Lake City, Utah. One of my original employees, Sid Siddeek, ran my office in Casper for some years.

During my years as a consultant, I greatly enjoyed what I was doing, although the company experienced the usual ups and downs. Generally, I felt Americans were straightforward and disinclined to stab you in the back. Now I also learned that there were exceptions to that belief.

I did a great deal of consultation for the University of Southern Colorado at Pueblo, which had tremendous problems with expansive soils. The Colorado State Department of Administration asked me to give a seminar on expansive soils at the university. The head of the state construction division attended and praised me as the most knowledgeable person on the subject in the United States.

A year later I was asked to do the soil investigation for a proposed nursing home in Florence, Colorado. Unfortunately, some foundation movement occurred shortly after occupation, mainly as a result of overwatering of the new landscaping. Without warning, the state of Colorado filed a suit against Chen and Associates that was initiated by the same person who had praised me so highly the year before.

Many experts were called as witnesses. Upon reviewing my work, all agreed that I had done nothing wrong and that my report was sound. The state then claimed that engineers from Colorado were biased and had testified to protect their fellow in-state engineers. The state went to Chicago and New York in search of outside opinions. Finally, it located an unlicensed architect, who claimed my report was contradictory and confusing. Based upon his evaluation, the state issued a letter to all concerned, saying "that serious questions have arisen as to the ability of Chen and Associates to perform in a first-class, substantial manner on a major construction project. Chen and Associates are advised that their work for the state will be closely scrutinized." I was hopping mad. Until then, I had believed that in the United States you were innocent until proven guilty. The state had taken the position that I was guilty until proven innocent.

The entire episode took two years to resolve. When the state finally realized it did not have a case, it dropped my name from the suit. During those two years I had spent more than $30,000 defending my work on a project that had an original fee of $2,000. I developed serious doubts about the U.S. legal system and the moral fiber of some government bureaucrats. In some ways I might as well have been back in wartime Chongqing adding chops to government paperwork.

Chapter Nineteen

Engineering in the Rockies

Of course, I also experienced moments of pride and delight in my accomplishments. Pikes Peak stands 14,110 feet high. Although it is not the highest peak in the Rocky Mountains, it is the best known.

The existing guest house had been constructed in 1940. It had developed serious foundation problems, so the owner had decided to build a new one. When I looked into the situation, I found that the old building had been damaged by permafrost.

To my knowledge, this was the only place in the United States that had a permafrost problem except for Alaska. When the foundation system intruded into the permafrost, rising temperatures resulted in frost heave. After a thorough study of the area, I designed a foundation system that had never been used previously: it isolated the foundation from the permafrost. The system worked, and the new guest house stood without any damage. Being the first engineer, to my knowledge, to tackle the problem of a foundation in permafrost gave me great personal satisfaction.

I was also asked to handle another difficult expansive soil problem. The site of the proposed community college in Colorado Springs had extremely explosive expansive soil. Most geotechnical engineers, as well as the state geologist, advised the administration not to build in such an unstable area. However, the school said that since the land had been donated, it could not afford to abandon the site. It was willing to spend additional money for a foundation system. I was called in to evaluate the possibility of using the site.

After extensive investigation, I concluded that the land had the most highly expansive soil I had ever seen. However, if the school was willing to pay for proper foundation work, we had a possibility for success. I decided on a drilled pier system, with the piers extending to a great depth, coupled with a deep trench system as a subdrain. The deep drain system was necessary to prevent future perched water from attacking the pier system.

To ensure that the trench was properly excavated and to take soil samples, I asked the excavator to lower me down to the bottom of the trench. While I was gathering the soil samples, without any warning the vertical wall of the trench collapsed. I was buried to the top of my head.

Luckily, the excavator saw what had happened. He jumped into the trench and freed my head so I could breathe. Gradually, the workers dug me out, and I escaped without injury. I gained a healthy respect for nature from that experience.

The college was built, and more than ten years have passed with no complaints about the foundation. In my experience with building construction, if the owner or the occupants do not complain, the building must be in good condition. As the old saying goes, no news is good news. People never call to congratulate you on a construction job well done, but they will call immediately if the slightest crack appears in the foundation, or if any foundation movement occurs. I felt proud that this project succeeded.

Everyone in Denver is familiar with Mile High Stadium, the home of the Denver Broncos. The stadium was originally built for minor-league baseball, and when the Broncos football team moved into it, the capacity was inadequate. To make the stadium suitable for both the Broncos and baseball, the people in charge wanted a movable stand of seats on the east side that could be set up for football and taken out for baseball. At that time only one other set of movable stands existed. It was in Hawaii and was much smaller than the proposed stand for the Broncos' fans.

Chen and Associates received the contract to design the foundation system. The specifications called for less than one-fourth of an inch differential in elevation to allow the stand to be easily moved. The existing east stand rested on soft fill-dirt, so we used a drilled pier system that extended down to bedrock. After more than a decade of use, the movable stand remains a complete success. Fans attending games in the stadium rarely realize they are sitting on a movable structure.

In thirty years of business, we did more than forty thousand projects. When we opened offices in San Antonio and Phoenix, we felt ready to handle work in most of the western states. Our projects ran the gamut of building foundations, from new designs to repair of existing structures. We worked on

water and sewage treatment plants
dams and reservoirs
transmission and microwave towers
mills and tailing impounds
coal-mining facilities

landslide assessments
slurry cutoff walls
stability analyses
tiebacks and retaining walls
dam foundations
docks and warehouses
cracked commercial and residential buildings
stadiums and arenas
airport runways and aprons
highway pavements and pavement stabilization
bridge foundations
water tanks and storage bins
bakery and meat-processing plants
various housing developments

Ultimately, Chen and Associates worked on almost every conceivable kind of construction project.

Additionally, we did the foundation systems for most of the high-rise buildings that now make up the Denver skyline, including Denver West, where we used a bearing pressure of eighty thousand pounds per square foot. We computed a six percent increase in pressure for each foot of penetration of the pier into the Denver blue shale. Such high pressures were regarded as dangerous by geotechnical engineers from the eastern United States, but the piers have stood.

We became the largest geotechnical company in the Rocky Mountain region, with more than two hundred people on the payroll. I believe we were also at the top technically. Our sales accelerated rapidly, from $1.5 million to more than $10 million per year. In 1980 *Engineering News Records* listed us as the 250th-largest engineering company in the country. When Colorado made a bid for the superconductor/supercollider facility, we participated on the design team.

At the time, I realized that for us to achieve national recognition, technological expertise was not enough. We had to become actively engaged in the international marketplace, in the local political arena, and in the leadership of engineering societies, publications, and academics. I could not do all of this myself. From 1978 on, I left the daily affairs in the hands of my associate Dick Hepworth and thereby freed myself from the routine chaos of business. This ultimately proved to be an excellent decision.

– *Chapter Twenty*
Political Involvement

I genuinely believe U.S. citizens have a responsibility to participate in the political arena. Most professional people I know consider politics to be a dirty business, and they refuse to bother. Politicians are well aware that they have passed laws concerning engineering matters without consulting any engineers, but those in the engineering community realized far too late that all they could do was curse the government after the fact.

From 1974 on, I continued my activities in professional societies as part of my involvement in politics. I believe I was the first engineer in the history of Colorado to serve as president of three major engineering societies — the Consulting Engineers Council, the Professional Engineers of Colorado, and the American Society of Civil Engineers. I served two years as senior vice president of the National Consulting Engineers Council, and during this time I was actively engaged in state legislation, academics, and high-tech research. One bonus of my outside activities was that they made our company the preeminent geotechnical organization in the region, with seven branch offices in the Rocky Mountain states.

I decided to become an active participant in the legislative sessions. Our lobbyist was good enough to inform me about which bills were of potential concern to engineers. I gave testimony on the strip-mining legislation in 1975, which led to my becoming a member of the Colorado Land Reclamation Board. During my two years on the board, I realized the importance of geotechnical engineering for land reclamation. I also discovered the great concern generated by the competition between the public sector and the universities. My involvement with the Land Reclamation Board gave me the opportunity to meet and to get to know the high-level administrators in the colleges and universities in Colorado. Thus, my business was also indirectly advertised.

Next, I became involved in the National Construction Industry Council and was also involved in construction circles. In 1974 I attended the soil-cement workshop sponsored by the Portland Cement Association, and I related my experiences with soil-cement construction in 1937 outside Xian. To my knowledge, that road was the first soil-cement road ever built.

Probably the most important legislation with which I was ever involved covered professional liability claims. In the thirty years since I had first examined the issue, the problem had gotten worse. Since 1960, professional liability claims in the United States have increased dramatically for engineering firms — from twelve and half claims per one hundred firms to twice that rate in 1975.

Engineers face multiple jeopardy, unlike doctors, dentists, product suppliers, and others. We have to deal with direct claims and third-party suits. On November 16, 1976, I testified before a house committee on behalf of the legislative committee of the Consulting Engineers Council. Among those present were Dale Tooley, then the district attorney, and J. D. McFarlane, the attorney general. I concluded: "The fate of the design profession is in your hands. If the present trend of liability lawsuits against engineers is not checked, the governing principle of private enterprise will diminish and eventually disappear. By passing reasonable legislation against nuisance suits, you are not only keeping alive the spirit of free enterprise, you are also protecting the public against unnecessary costs in engineering projects."

I expected to generate a lot of debate among the legal associations, but they remained quiet, whereas the insurance companies reacted vigorously. The leading professional liability insurance company, Victor O. Schinnerer and Company, responded through its agent: "It is possible for engineers to reduce liability claims simply by improving the quality control procedure within their firms. Protecting against liability claims is often nothing more than exercising good business practices and old-fashioned common sense."

Actually, I felt insurance companies and lawyers worked hand in glove on this issue. After all, the more claims that were filed against an engineer, the higher all engineers' premiums became. In fact, this problem was not limited to engineering; during the same years nuisance malpractice suits against the medical profession also increased, suggesting that the legal profession, not just engineers and doctors, were behind this trend.

Naturally, not every lawyer joined the bandwagon. One fair-minded attorney, when discussing the tort of unreasonable litigation and how to stop legalized extortion, wrote that any defendant who has been sued without good cause, and any defendant who has suffered the abuse of excessive discovery, should fight back. If the defendant is injured, then his or her insurer has also been injured and should also have a cause for action.

After three years of continual fighting, the legislators of the state of Colorado finally passed a watered-down bill in 1977 that curbed some frivolous and groundless suits by awarding the attorney's fees to the defendant. Unfortunately, the bill that was passed into law in 1977 has done little to curb the onslaught of frivolous suits. Presently, engineers spend more time and energy in the reports they write trying to avoid suits than in trying to do good engineering. Ninety percent of all cases are settled out of court, which is exactly what the plaintiffs' lawyers and the insurance companies are seeking. Current statistics paint a grim picture. Consulting firms pay four and a half percent of their annual billings for liability insurance, and this figure is probably low. Even if the claim is frivolous, the cost of defending against it will be exorbitant. In sum, figuring in the money lost in wasted billable hours and the cost of the defense, these suits cost a firm more than $20,000 per claim. Naturally, these costs must be passed on to future customers.

For many years I testified as an expert witness on engineering matters. Of the more than one hundred times I have testified, I have not testified against my fellow engineers. I am appalled by the fact that some engineers testify against their fellows for a small fee. A very strong ethic should be required for anyone acting as an expert witness. We need to speak to the truth of the matter and not just say what the lawyer wants us to for money. Our legal system hinges in part on the effectiveness of expert testimony, and we can help to improve the system through rigorous scrutiny of our own ethical standards.

I was also successful in helping achieve the passage of a bill that sets the statute of limitation for civil action in Colorado at ten years. Finally, Colorado also abolished joint and several liabilities, ending engineers' multiple jeopardy. In my opinion, these moves should have been made long before.

In 1977 I was appointed by Governor Richard Lamm of Colorado to the State Board of Registration for Professional Engineers and Professional Land Surveyors. The board's task is to police engineers and reduce malpractice. The board is under the jurisdiction of the Department of Regulatory Agencies (DORA).

Before long DORA wanted to abolish the position of executive secretary of the registration board and claimed that its criteria for registration were too strict. I looked into the provisions of the Colorado statutes pertaining to the regulatory board and found that DORA's action was clearly illegal.

This was the beginning of a long legal battle between the engineers and the state bureaucrats. I was pleased to see that for once all of the engineering societies were united in a common effort. We retained an attorney to counter the state's legal claim. It was apparent that someone within the state did not like our board's criteria for membership and claimed that our qualifications were so strict that minorities were excluded.

I could clearly see that in the future DORA, to boost its authority, would keep the board under its thumb. Future board members would only be acting as a rubber stamp for DORA. The bureaucracy would grow, efficiency would drop, and complaints against engineers would be ignored — all in the name of "protecting the health, welfare, and safety of the public."

In 1981 the state offered to withdraw its proposed legislation and to seek some kind of compromise. I considered that we had won the case. I wrote to the governor that I would not seek reappointment to the board.

I also became involved in the Colorado Association of Commerce and Industry (CACI). Members of the association included Marathon Oil, Interstate Bank, Hewlett Packard, First Interstate Bank, and IBM. I also chaired the interface committee of the Higher Education Council (HEC). The committee's function was to coordinate activities with local Chambers of Commerce, professional societies, political action committees, and all other organizations with any interest in higher education. The appointment gave me the opportunity to have close contact with the three leading state universities: Colorado State University, the University of Colorado, and the School of Mines. I was amazed to learn how little Colorado legislative bodies knew about education and how the funding for engineering education was distributed.

The engineering societies presented a position paper on the funding of engineering education to the state legislature, with facts and figures. Unfortunately, the efforts of my committee had no impact on the legislature. It appeared that the bureaucrats paid little attention to the need for sound engineering until an accident forced them to notice. The failure of the Lawn Lake Dam garnered a great deal of public attention. I gave a talk and held a workshop on dam safety and liability.

I was informed at one point that since I owned more than fifty percent of the stock of Chen and Associates, the company should qualify as a minority business enterprise. According to the U.S. government, minority groups include Asian Americans. When I brought the matter up with the state, I was disqualified because my gross income was too large. I didn't argue, because I really don't agree in principle with the idea of minority business enterprises, although my failure to press the issue did lose us several multimillion dollar contracts. I feel the important issue should be whether the company in question can do the work rather than who owns the company.

Looking back on my political involvement, I experienced moments of great elation and gratification. I reflected then on being a foreigner who was achieving his dreams in this land of opportunity. On the other hand, I sometimes felt great frustration and doubted that the legal hassles were worthwhile. I often wished I shared my wife's faith in God, which helps her face her troubles with confidence.

– Chapter Twenty-One
Saudi Arabia and Liberia

In 1975, with the growth of the overseas market in construction, Chen and Associates decided to join three other leading engineering firms in Denver to form an industrial consulting group. With this group we could handle any industrial project around the world.

Saudi Arabia became our primary target. The growth in the country's oil income had produced a boom in Saudi construction. Nations around the world were rushing to establish offices in Riyadh, hoping to share in the oil wealth.

South Korea was the most aggressive country in this competition. The country not only sent good engineers but also backed them up with political connections to high-level Saudi bureaucrats. Ultimate political power was wielded by more than a dozen Saudi princes whose approval and support determined every business venture.

We tried to do business as we would in the United States. From Denver, our group entered a bid on a sports complex to be built in Riyadh. We did extensive research for the project, and when the bids opened in public we were the lowest bidder. We rejoiced, but only briefly, because the prince in charge of the project awarded the job to South Koreans. His decision was final, and we were bitterly disappointed.

I told our group that we should no longer try to conduct business in Saudi Arabia from Denver. If we expected to accomplish anything, we would have to have an office in Saudi Arabia. At the end of 1976, I had many discussions with a representative of Prince Turki, a man named Nasser Al-Anbar.

Although Al-Anbar was a typical salesman in some ways, he did not have the smooth-talking style of a U.S. salesperson or any knowledge of construction and engineering. I proposed to him that our group would establish an office in Riyadh that could handle all geotechnical work in the kingdom, from fieldwork through laboratory testing to issuing final reports. The proposed initial cost neared $30,000. I anticipated that at the end of five years we would generate a gross income of $400,000. Prince Turki would own fifty percent of the company. This last figure impressed Al-Anbar.

In early 1977 I made a trip to Riyadh and took with me a multilingual young man from Beirut. He was a structural engineer who was fluent in Arabic, French, and English. A close friend of his from the University of Colorado now held an important post in Riyadh under Prince Turki. Connections of this sort seemed to be more important than I had realized.

When we arrived, the heat in the Arabian desert was unbearable. By noon the temperature was well over a hundred degrees, and all activity came to a halt. I couldn't imagine how any business could be done under the scorching sun.

In the city of Riyadh we saw new buildings, new streets, and new hotels, but the systems for water, telephones, and sewage were still primitive. In fact, this helped me in a minor way: since I could not speak a word of Arabic and could not read the street signs, I had a hard time finding my way back to my hotel. Finally, I recognized an open sewer near the hotel and found my way back by the smell.

As I became familiar with Riyadh, I saw two preponderant qualities in the city. Most of the city had been built very recently, and all of the oil money had made cost irrelevant. Tremendous waste lay everywhere, not only with construction projects but throughout the entire administrative system. I saw Mercedes Benzes abandoned by the side of the road, lacking parts or mechanics who could fix them. Apparently it was cheaper to buy a new car than to fix the old one.

At the same time, I became better acquainted with the Arabs. Needless to say, the Arabs conducted their business in a completely different style from the way it was conducted in the United States. Contracts and verbal agreements had no meaning. The overriding element was the whim of the prince in charge. The South Koreans understood this, and consequently they had a very successful enterprise in Saudi Arabia.

A nod from the appropriate prince meant you got the contract you sought. The Arab architects and engineers were not concerned about the workmanship of the contractors, even though the contractors were paid roughly three times a normal fee. As a consequence, the work of the Arab architects and engineers was uniformly substandard. The electrical outlet plates in my room did not even cover the hole in the wall, which gave mice and insects easy access. In recent years I have heard that much of this has changed and that a modern Riyadh has blossomed in the desert.

I found the gold market particularly interesting. It consisted of small booths on both sides of the street that had no covers, doors, or locks of any kind. Anyone could easily reach over the counter and grab a handful of gold. I guessed that each booth contained about a million dollars in gold.

At the time I visited, one booth owner walked away from his booth, leaving it unattended. In the United States strangers would have looted it in an instant. I found out later that if anyone was caught stealing, it meant the loss of a hand. No wonder the booths could be left without worry.

Early one morning I was awakened by a commotion in the street. Everyone was rushing to the market square. A hotel staffer told me they were going to witness a public hanging. I also learned that if a woman was caught being unfaithful to her husband, she would be stoned to death in the public square. Apparently, ancient biblical practices remained in force. I had to wonder why the United States, which makes such a big issue of human rights in some countries, closes its eyes to places like Saudi Arabia. Is it because we need that country's oil?

At last, I received an invitation from Prince Turki to dine at his palace. When I arrived I was deeply impressed by the opulence of the furnishings and the elegance of the dinner. We got along well, and the prince accepted my suggestion to open an office in Riyadh, of which he would own fifty percent. My trip to his country had been successful.

Prince Turki was well informed about events in Communist China. He asked me questions that demonstrated a knowledge of Chinese affairs few people had. I suspect that when Saudi Arabia resumed diplomatic relations with China in later years, Prince Turki may have played a central role. Each morning, when I was awakened by the prayer caller from the mosque, the sound reminded me of my wartime days among the Chinese Muslims in Qinghai. The sound was the same even though the language was Arabic instead of Chinese.

After my experience with Muslims in Qinghai and what I had now seen of Muslims in other countries, I concluded that it was wrong for Christians to call members of other religions "nonbelievers." I believe there is only one God, whether we elect to call that god Allah, Buddha, or Zeus. Humans seem to have a narrow perspective when it comes to religion. Everyone believes his or her view is the only view. All of the religious leaders in history have espoused a narrow view. A possible exception is Buddhism, which carries the implication that more exists beyond the Buddha. Confucius refused to discuss the subject when queried by his followers, simply saying that he did not have the answers. Many of the major wars in history began with a religious argument. I have to wonder if it is possible for Christians, Jews, Muslims, and

Hindus to coexist. Maybe someday all religions can be fused together in a great metaphysical melting pot.

My own experiences clearly indicated to me that there is an almighty God. This God may exist in the form of a human being or simply as a mathematical equation that holds the key to all other equations. We do not know. As a Chinese philosopher once said, "Our lifespan is limited, but the knowledge of the universe is limitless. It is a hopeless effort to try to measure the limitless with our limited lives."

After I had returned to Denver I talked with my former employee from Iraq, Sid Siddeek, and asked him if he would be willing to open an office in Riyadh. Since he was from the Middle East and he understood my business, he was a perfect choice. He agreed. We ordered a drill rig from the Central Mining Company and some basic soil-testing equipment from Soil Test. In April 1977 Siddeek left for Riyadh. Since at the time communication between Denver and Riyadh was difficult and slow, he would have sole responsibility for the entire venture.

At the end of March 1978 I visited Saudi Arabia for the second time to check on my office in Riyadh after a year of operation. I had some trouble with a connecting flight in London and finally arrived in Riyadh at 2:00 A.M. Riyadh was like New York City on New Year's Eve, packed with fortune hunters, tourists, and immigrants. I was lucky to find a place to stay for the night.

Siddeek took me to his office the following morning. It was adequate for the task but no more. He told me stories about the way of doing business in the Arab world. Most of his troubles came not from the local people but from the Corps of Engineers. They were working on a multimillion-dollar airport, and gaining access to the inner circle of the corps took special dealing. Fortunately, Siddeek was knowledgeable in this art. I assured him that as far as I was concerned, all he had to do the first year was break even.

The next day we gave a reception at the office. We invited all of the appropriate dignitaries and provided a large punch bowl. The punch contained two bottles of whiskey I had purchased in the London airport. At the time, I didn't realize that drinking alcohol was illegal in Arabia and that even possessing the beverage was forbidden. Yet nobody seemed to notice the alcohol in the punch, and nothing was mentioned.

The following day I gave a talk on expansive soils at the university, with more than a hundred students in attendance. As I spoke, I noticed that each student was served a free beverage. Later, I also learned that being a student at the university makes one a member of the elite. Not only was everything free, but the graduates were also guaranteed government jobs.

I left Riyadh feeling relatively pleased with my contacts and with the progress of the office. On my way home I had a stop in Liberia. I had never been there, but I had heard about the country because my brother, Dr. Tai Chu Chen, had been the ambassador to Liberia from Taiwan from 1962 to 1966. The project I had to discuss was the improvement of the 155-mile Robertsfield Road. From Riyadh, I flew to the Liberian capital, Monrovia, by way of London. Mr. Bynum, my local contact, met me at the airport, but my timing was bad; President Carter was scheduled to arrive shortly, and the only decent hotel in Monrovia, the Ducar Palace, was booked solid.

I wound up in a place called the Holiday Inn for the night. This Holiday Inn had no relationship with the U.S. hotel chain. It was dirty and smelly, and worst of all, they had not changed the sheets between guests. I complained to the manager and got no satisfaction. As a result, I slept with my clothes on.

Luckily, I found an opening at the Ducar Palace the next day, and I moved immediately. However, no business could be conducted while President Carter was in town, because the entire city was in jubilee. Finally, after Carter left I managed to get in touch with Colonel Railey, who was in charge of President Tolbert's business. Our proposal for the road was duly submitted to the president through Colonel Railey. After my subsequent visit with President Tolbert, I discovered that the highway project was more of a personal project for the president than a public work to improve Liberia. That told me something about what kind of government ran the country.

During the course of the Robertsfield Road project, I had the opportunity to see the real Liberia, the one beyond the boundaries of Monrovia. I thought I had seen poverty in northwestern China, but I learned better. Here in the hinterland, the Liberians were naked and lived in mud huts, eating out of whatever garbage cans they could find. As a result, they died like flies. I was not surprised to hear recently that Liberia was experiencing a revolution and that both President Tolbert and Colonel Railey had been executed.

I left Monrovia in April 1978. A customs officer noticed my German camera, which I had purchased in Berlin many years earlier. He demanded that I pay an export duty. After my earlier experiences with corruption, I knew I had no choice, so I handed over a twenty-dollar bill. I was reminded of how my brother had described Liberia during his ambassadorial tenure. It seemed nothing had changed.

– Chapter Twenty-Two
China Revisited

After President Nixon's visit to the People's Republic of China in 1972, the icy relationship between China and the United States gradually thawed. China finally established a liaison office in Washington. By 1978 I understood it was possible for a U.S. citizen to apply for a visa to visit China, either as a tourist or to visit family members. I decided to return to China after an absence of almost thirty years.

On December 17, 1978, I left Los Angeles and flew directly to Beijing with my wife, Edna, my daughter, Yvonne, and my son, Tyrone. We arrived at the Beijing airport expecting to find our own way into the city, but we were surprised to be greeted by the Chinese travel service, whose representatives took us to the hotel and gave us a banquet. The travel service also asked me for a list of people I would like to see and said they would arrange for them to come to my hotel. I was shocked to be treated like a VIP.

After the death of Mao Zedong that year, President Hua Guofeng had taken charge of the country. We were able to see first-hand what the interior of Communist China had become immediately after the horrors inflicted on the country by the cultural revolution. The China I saw was a far different place from the one I had left thirty years before. The best way I could describe the new China was that it was very uniform.

Everyone appeared to have been cast in the same mold. They dressed alike, expressed the same opinions about the regime, rode the same kind of bicycles, and went to work at the same hours. I was amazed that the Communists had been able to transform what I remembered as the highly individualistic Chinese people into an apparent mass of robots.

On the street, people immediately recognized us as foreigners by our clothes, and they flocked around us asking questions. I took pictures of them with my Polaroid camera. They were startled by the instantaneous prints. The people were very grateful to have the pictures, but unfortunately I had only one pack of film.

I could not help noticing that this city of millions of people had no daily newspapers. People got their news by reading wall posters, which were handwritten in striking colors. The wall posters drew crowds who gathered what news they could of domestic or foreign events.

At lunch one day, two gray-haired gentlemen approached me and asked whether I recognized them. I hadn't the foggiest idea who they were. When they told me their names, I realized that they had been my assistant chief engineers on the Qinghai-Tibet Highway.

We talked about old times for hours, and I was glad to hear that both of them had important positions in the Ministry of Communication. Both were scheduled for retirement in a few years. Some of my fellow engineers had died during the cultural revolution. These two men related to me, without reservation, how they had suffered during those years. China would be a far different place had it not suffered the setbacks imposed by the Gang of Four.

I was especially interested, of course, in engineering projects. My former assistants told me that an incredible amount of new construction had been completed in the past thirty years in spite of the cultural revolution. The Tibetan Highway had been extended from Yushu to Lhasa and was now paved. Railroad lines wound through the mountains in Yunnan, where we had thought they were impossible to build. Hundreds of new flood control and irrigation projects were in operation. Lanzhou, the insignificant, remote city from which I had begun the Lanzhou-Sichuan Highway, had become a major industrial city, the hub of the petrochemical industry and the site of China's nuclear facility. In general, the lives of ordinary people had improved dramatically. The improvements were probably not obvious to the eyes of most foreigners, but for me, having seen China before the Communist regime, the changes were beyond my wildest dreams, especially in northwest China. I think that had it not been for the setbacks of the eight years of the cultural revolution, China would have already emerged as one of the industrial powers of Asia.

On one cold, snowy afternoon, I took my family sightseeing along a busy street in Beijing. We dropped into an eatery that was packed with local patrons and ordered all of the delicacies on the menu. This excellent lunch cost us the equivalent of about two U.S. dollars. We paid the bill in local currency, with a little extra, and walked out well satisfied.

When we were about a block away, the waitress caught up with us and gave back the change, saying, "We do not accept tips." Still wanting to express our appreciation, I gave her a used felt-tipped pen, which she happily accepted.

I had a difficult time finding my childhood house in Beijing. We had lived in the house more than sixty years earlier, when I was a teenager. I remembered a huge house with a courtyard where I flew kites and played games. The house was still there; nothing had changed, except I now found it a dingy little place. The present owner was very hospitable, and we talked at length. The house was over a hundred years old and had changed hands several times, although very little had been done in the way of repairs or additions. I left feeling strangely disappointed. Maybe I should have left the memory of my childhood intact instead of poking holes in my dreams.

After sightseeing at the Great Wall, the Summer Palace, Beihai Park, the Imperial Palace, and the other tourist attractions, we left Beijing and flew to Shanghai. All of our travel arrangements were handled by the China Travel Service, and everything went smoothly.

First on my agenda in Shanghai I paid a visit to my former boss, Zhao Zukang. He was nearing ninety years of age, but he was still in excellent health, especially considering that he had suffered from tuberculosis as a young man and had not been expected to live very long. I also had the opportunity to meet Dr. Zhu again, the man who had become separated from our group when we fled from the Japanese on the Burma Road and met me in Baoshan days later.

Zhao Zukang arranged a banquet in my honor with friends from years ago, and he asked me to address the Society of Chinese Civil Engineers. More than five hundred top engineers turned out, and I met many old friends. I spoke for more than two hours and enjoyed their company very much.

After Shanghai, we went on to Hangzhou. This city boasts the most scenic spot in China, the West Lake. The local expression is that "there is paradise in heaven, and there is Hangzhou on earth." The purpose of my visit was to pay homage to the tomb of my father, who had been buried there as a martyr. Back when I was young, a stele (tablet) had been erected over his tomb, with an inscription from Dr. Sun Yixian. I felt proud to learn that the tomb was held in high regard by Chinese tourists.

When I reached the burial site, I found that both the stele and the tomb were gone. Nothing marked the burial site. The local people told me the Gang of Four had had the site demolished during the cultural revolution. The present regime has done a great deal of restoration work on monuments to the early martyrs of the revolution against the Manzhou and the warlords. I hope my father's tomb will soon be restored.

West Lake retains the beauty and tranquillity it has possessed throughout the centuries. I also enjoyed sipping the best tea in China. I was convinced that it was not only the tea that yielded the great flavor but also the water from the famous Dragon Well.

Our last stop in China was Fuzhou, where my mother had spent the last few years of her life. She died at age ninety when I was in Denver; it was heartbreaking to see her bedroom and her few remaining possessions. I remembered again that the worst failure of filial loyalty I could have committed was to be absent during her death. Now all I could do was donate her meager estate to the state and take a few articles as mementos.

Even so, this was the most pleasant trip I had taken for years, since I had no particular obligations to meet. I could go wherever I pleased, and we stopped anywhere we liked. Everyplace I visited brought back profound memories.

Back in Denver I was amazed at the number of requests I had to speak about my trip to China. I wrote special feature articles for local papers and appeared on television. The four hundred slides I had taken were very popular.

My second return to China occurred at the request of the State Capital Construction Commission of the People's Republic of China. Our group consisted of a mining engineer, a port and harbor specialist, and civil and geotechnical engineers. We departed from Los Angeles November 27, 1979, and arrived in Hong Kong the next day. From there, we traveled to Guangzhou (Canton) by rail.

Guangzhou was the birthplace of the Chinese revolt against the Manzhou Dynasty, and since it has had more contact with the West, it is considered to be the most westernized city in China. In fact, most of the immigrants to the United States came from Guangzhou or at least from Guangdong Province. Our contacts told us that during the cultural revolution Guangzhou was heavily damaged and greatly demoralized.

Today, the city is booming. Half of Chinese foreign trade emanates from Guangzhou, and it is called the fifth of the "small dragons of Asia," along with Hong Kong, Singapore, Taiwan, and South Korea. In Guangzhou I tasted fresh litchi nuts again and found them as delicious as ever. They reminded me of Yang Kuefei and her tomb at Mawei. She was said to have been so fond of litchi nuts that the emperor had them transported from Guangzhou to Xian by pony express just for her. The last batch arrived at the same time as her hanging as a result of palace intrigue, and when the emperor saw the fruit he could not contain his grief. Today, this story is still told all over Guangzhou.

We flew to Beijing, where we were greeted by Chinese officials. We gave a talk on pollution control, a topic they certainly needed to know more about, because the air pollution was so bad it clouded the streets like a dense fog. I gave a talk at the Coal Industry Research Institute on expansive soils and mine shaft design. I met Liang Weisen there, and I was so impressed with his geotechnical knowledge that I invited him to work for my company in Denver. He later settled in the United States and arranged to have his wife and daughter join him.

Our team gave several talks, including one on feasibility studies, something the Chinese had never considered. Most projects in China were started without any consideration of economic, environmental, or technical impacts. For example, the Chinese spent billions of dollars establishing a steel mill near Shanghai on the advice of Japanese consultants. Later, they found out that the site was unsuitable, and the mill stood idle. The Japanese didn't care, because they had already taken their profit from the sale of consultation and equipment.

Our Beijing trip concluded with a dinner at the Peking Duck Palace. The duck was definitely better than any I had eaten in the San Francisco or New York Chinatowns. However, it still was not as good as I remembered from my childhood.

In Shanghai the team gave a talk at the Grand Theater. All fifteen hundred seats were filled. My concluding remarks, in Chinese, were probably the only part of the talks the audience understood. The rest of our stay in Shanghai was devoted to the Shanghai Tunnel Construction Company, the Shanghai Geological Bureau, the Shanghai Industrial Design Bureau, the Shanghai Construction Science Research Institute, and the Shanghai Foundation Construction Company. We found the Chinese engineers were highly competent, and they had a good understanding of current technology.

In Hangzhou, while the team was taking in some of the most interesting scenery in China, I was asked to be the keynote speaker for the national conference of the Chinese Society of Soil Mechanics. It took me back to 1941 when I was in Nanjing, where this society had been founded. The chair of the society then was Dr. Mao Yisen, a world-renowned engineer, and I was the society's secretary. I was pleasantly surprised to find that after forty years, the society was still in existence and was holding its first meeting since the cultural revolution.

The two hundred top geotechnical engineers from all over China attended. I saw familiar faces from the time I had been in Berlin. Most of the older engineers were now teaching in various universities. I gave a two-hour talk on expansive soils. For me, attending the conference was a memorable event, and I considered the trip resoundingly worthwhile.

Altogether, we had given nine lectures and been wined and dined at nine banquets. Our work had been substantive and satisfying. We concluded our mission and returned home December 14, 1979.

In the following years, communication between China and the United States increased by leaps and bounds. I accepted the role of chief negotiator for all Chinese affairs for the American Consulting Engineers Council, the largest organization of engineers in private practice in the United States.

China sent several missions to the United States, the most important of which were the Three Gorge Project, under the auspices of the U.S. Trade and Development Program, and the State Capital Construction Commission. This commission consisted of six people with both technical and management experience, as well as familiarity with governmental structure. The party's twenty-eight-day tour included visits to San Francisco, Baltimore, Washington, New York City, Denver, and Houston, where NASA's Space Center was the focus of attention. I accompanied the party for the entire tour, which wound up in Denver.

During the entire month of June, I did not set foot in my office. By the time the month ended I had had enough of giving speeches, doing translations, and attending banquets. I saw our visitors off with a sense of relief in July 1981.

The State Capital Construction Commission of the People's Republic of China signed a tentative memorandum of understanding with the American Consulting Engineers Council. However, I had the impression that the visiting Chinese did not really understand what a consulting engineer is. This was no real surprise, since for the last half century everything in China had been controlled by the central administration. No consideration had been given to the cost of a project, its feasibility, or the potential profits. The commission was quite impressed with our efficiency at Chen and Associates. In China, to do the work of a small geotechnical company like ours, which had two hundred people, they would need about a thousand people because of the inefficiencies.

At the same time, I was impressed with the decisionmaking process in China. When the top officials gave the go-ahead for a project, it was begun at once. A project like the Three Gorge Dam could never be built in the United States. The environmental impact study would take ten years, and hundreds of lawsuits would be filed by the local people as heated debates raged in the House and the Senate. Insurance companies, labor unions, and conservationists would all eventually have a hand in the process.

This was not the case around 1934. I was in Las Vegas during the construction of the Hoover Dam, when the Bureau of Reclamation was the final authority on almost every detail. Today, the bureau has only a fraction of its former authority.

The Consulting Engineers Council sent a return delegation to the State Capital Construction Commission. The trip was scheduled to coincide with the annual conference of the International Federation of Consulting Engineers, to be held in Singapore during June 1982. I arranged to present a paper on corporate practice at the conference. This time, my wife, Edna, and my daughter, Yvonne, accompanied me on the trip to Singapore and China.

This trip gave me another chance to return to cities in China I had known in my youth. After my talk in Singapore, which went well, we flew to Hong Kong. I had not been back since we had left in 1957.

All of the small buildings in Hong Kong seemed to have been replaced with high-rises. The twenty-four-story building I had designed in 1950 had been replaced with a structure twice its height. The ferry connecting the Kowloon Peninsula with Hong Kong Island had been supplemented by a tunnel under Hong Kong Bay. For me, who had lived in Hong Kong for seven years, the place was barely recognizable. Hong Kong was booming in every respect.

We arrived in Beijing June 25, 1982. The Consulting Engineers Council expected to enter into an agreement with the Capital Construction Division of the State Economic Commission. I did most of the negotiating, and I found communication to be difficult, not because of language but because the Chinese still did not really understand the role of a consultant. Because of their limited experience with private enterprise, they did not understand why a consultant would charge a high fee. I could not convince them that a consultant could save them money in both design and construction later in the project.

During our various conferences I met many former colleagues who now held high positions in various ministries. I felt these renewed acquaintances helped in our negotiations. Meanwhile, Zhao Ziyang had replaced Hua Guofeng as premier. He sent the vice premier to greet us at a banquet where we had expected the premier to be present, a sign that he was not very interested in our negotiations.

To sum up negotiating in China, the word is "patience." We in the West will not harvest results in China in a short time. The seeds must be planted now, or nothing will come in the future. Years from now the American Consulting Engineering Council will probably thank these delegates for their earnest attempts.

While the other delegates toured the Great Wall and the Summer Palace, I visited the Yong He Palace, or Lama's Temple. According to legend, Emperor Yong Zheng, who ruled for only twelve years (1723–1735), was assassinated in the Yong He Palace by a young girl avenging the death of her father at the hands of the emperor. The temple is also known for the sacred images of the happy Buddhas, in which the god and goddess are shown in various positions of sexual intercourse.

From Beijing, the delegation flew to Xian. I eagerly anticipated my return to Xian after fifty years and was anxious to see what changes had occurred. The difference amazed me. Trees now lined the streets, vegetable gardens could be seen everywhere, and apartment buildings had replaced the cave dwellings. Most notably to me, the Xian-Lanzhou Highway, where the soil I had stabilized with cement had been cut to pieces by the iron wheels of horse-drawn carts, had been paved with asphalt.

My wife, who is a devout Christian, wanted to visit the Stone Stele Museum in Xian. The stone steles date back to the time of Confucius and relate historical events in China. One stele from the Tang Dynasty records the beginning of Christianity in China, manifested in the Nestorian (Jingjiao) sect.

Everyone in our group was eager to see the museum of the Qin Dynasty terra-cotta warriors and horses, where the first emperor was buried. Pit number one was discovered in 1974. The wooden and masonry roof had burned and collapsed, crushing most of the figures, but the pieces were intact and restorable. Six thousand pottery soldiers and chariots pulled by six horses stood there. Pits numbers two and three contained fourteen hundred figures, but no restoration work had been done. Originally, the warriors and horses were painted in bright colors, but the colors had faded in their twenty-two hundred years underground. The pits under excavation are only a small part of the peripheral area of the mausoleum.

After my long absence from China, my only living relative was my wife's sister, Ai Feng. We went to visit her and her husband in Tienjin, where I had gone to college at the University of Nankai the first time I had left home as a young man. Ai Feng was a popular medical doctor, who had been trained in the United States and was well-known all over north China. During the cultural revolution, Premier Zhou Enlai had protected her from persecution by the Gang of Four.

In fact, during the tumultuous years of the cultural revolution, Zhou saved many Chinese from persecution. Without Zhou's intervention, the outcome of the revolution could have been far worse than it was. Although Mao was respected during those years, Zhou has been called the leader for whom people had genuine affection.

In China, younger people with more liberal views are more frequently in charge of the government. Recent developments indicate to me that China is becoming an industrial nation, with a projected twelve percent annual growth rate. China could well emerge as the world's top industrial nation, given the country's vast natural resources, coupled with its technical knowledge.

In May 1984 I made my fifth return trip to China. The trip was taken at the invitation of the Communication Survey Design Bureau in Hunan. With me were Curtis Seely of Chen and Associates and Robert Janowski of Dravo Engineering Inc.

We traveled from Guangzhou to Shantou in a hired cab. This was the first time I had traveled by highway in the new China. Road conditions were comparable to a second-class county road in the United States. We had to dodge pedestrian and bicycle traffic in the two-lane road with no shoulders. At Shantou we inspected proposed hotel sites. We collected enough data to write a fifty-page report on feasibility when we returned to Denver.

We arrived at Changsha and were greeted by the vice governor of Hunan Province. I gave several technical lectures and signed a tentative agreement on the scientific and technical exchange memorandum of understanding among the Communication Survey Design Bureau, Hunan, and Chen and Associates in Denver.

Anyone traveling to Changsha should not miss a visit to the three Han tombs at Mawangdui in the eastern suburbs of Changsha. The first tomb, excavated in 1974, disclosed the body of a woman in remarkably good condition. The brownish-yellow skin is smooth and resilient, and the facial features are fairly distinct. It is generally agreed that the tightly sealed chamber and the many layers of charcoal and white clay covering the multiple coffins stacked one inside the other account for the extraordinary preservation of the body, which had been buried for more than twenty-one hundred years. I have seen the Egyptian mummies in the British Museum in London, and although they are much older, they are like minerals; they are nothing like this body with its lifelike features. I have heard that the body has begun to deteriorate, although it is stored in a vacuum chamber.

The second and third tombs were discovered a year later, and they enable scholars to identify and date the three burials. Tomb number three contained a remarkable group of ancient texts on wooden and bamboo strips and on silk. Preliminary research indicates that the tombs date to 168 B.C. and belong to the house of Doi. The contents of the three tombs are so extensive that it will probably take many years before the research is finished.

At the invitation of the Yangzie Valley Planning Office, Ministry of Hydraulics and Power, we left Changsha and headed for Wuhan. On May 28 we toured the new bridge across the Yangzie. The total length of the bridge is 3,808 feet. We also visited the soil-testing laboratory of the planning office. I usually judge the activity of a laboratory by its neatness. The laboratory was far too clean to have seen much activity.

The next day we took the train from Wuhan to Chongqing.The railroad line was very impressive. Most of the time the train was in tunnels. There were so many that the attendant did not even bother to turn off the lights whenever we emerged for a time. As a highway engineer, I could well imagine the extreme difficulties faced during the construction of the line.

While I was in Chongqing I tried to locate the apartment our family used to live in, but I could not find it. Then I tried to find the clubhouse where we were married. It used to be one of the elite places in Chongqing. Now, gone were the large bay windows, the beautiful waxed dancing floor, and the richly decorated walls. The place had been divided into apartments and was occupied by a number of families.

We left Chongqing on the *Red Flag* tourist ship and tried to absorb as much as we could of the scenery of the Three Gorges. The ship sailed too fast in the swift downstream current to allow us to see the sights. To fully enjoy the scenery, it is advisable to travel upstream.

From Badicheng to Nanjinguan, the Yangzie cuts through three mountain ranges, forming the celebrated Three Gorges — Qutang, Wuxia, and Xiling. The length of the gorges is about 130 miles. The jagged mountains and steep cliffs rear out of the water high above the rapids and reefs. The swift, roaring water and cloud-wreathed mountains present breathtaking views. Qutang Gorge, at it narrowest point, is less than 330 feet wide. The river churns its way through a wall of cliffs that hem it in on both sides. Travelers through this gorge may be reminded of the ancient description "Peaks pierce the sky, we course through a deep passage beneath."

Wuxia Gorge, the middle section of the three gorges, is flanked by grotesque peaks. The limestone on both banks has been sculpted into surrealistic forms. Xiling Gorge is the longest of the three gorges. Its stone corridor has sheer walls hundreds of feet high, and the river is marked by hazardous rapids and perilous whirlpools.

We arrived at Yichang in the afternoon, and the following day we took a chartered boat to Sandouping, the future site of the Three Gorge Dam. The location of the dam has been studied intensively for years. Geologists have reported that from Yibin to Yichang, the river bedrock consists primarily of shale, sandstone, and limestone. Only at Sandouping did they find solid granite, an ideal foundation for the huge dam. I was impressed by the core drilling operation. The diameters of the cores were three feet, and the cores extended to a depth of more than a hundred feet. The cores were laid out on the ground for detailed inspection. The engineers were interested in locating any seams or faults.

Near Yichang, about a mile from Nanjinguan, is the newly completed Gezhou Dam. The engineers took great pride in successfully diverting the river during construction. The diversion technique will undoubtedly also be used for the construction of the Three Gorge Dam.

Throughout history, Yangzie floods have caused serious damage. Over a two-thousand-year period, massive flood damage has occurred an average of about once every ten years. The economic loss from the flooding cannot even be estimated, but it has surely far exceeded the cost of building the dam. The dam will generate enormous amounts of hydroelectric power, improve navigation, and play a key role in the utilization of water resources for all of China. The dam will also help to realize the goal of sending water from the south to the north.

On the other side of the coin is the fact that the construction of the dam will destroy the historic scenery of the gorges. Unlike the Aswam Dam on the Nile, the flooded scenic landmarks cannot be salvaged. The construction means the relocation of at least half a million homes upstream. A comparison of the Three Gorge Project with some of the world's largest existing dams will give the reader some idea of the magnitude of the undertaking.

Three Gorge Dam

Height	640 feet
Length	7,000 feet
Concrete	29.2 million cubic yards
Spillway	2.75 million cubic yards
Power house	25,000 million watts

Itapu Dam, Brazil

Height	643 feet
Length	3,490 feet
Concrete	6.6 million cubic yards
Spillway	2.2 million cubic yards
Power house	12,600 million watts

Guri Dam, Venezuela
Height   631 feet
Length   4,336 feet
Concrete   6.6 million cubic yards
Spillway   1.2 million cubic yards
Power house   10,000 million watts

Grand Coulee Dam, Washington state, United States
Height   550 feet
Length   3,887 feet
Concrete   9.7 million cubic yards
Spillway   1.0 million cubic yards

The Three Gorge Dam thus ranks easily as the largest in the world. Intensive debate on the feasibility and even on the desirability of the dam has been going on for nearly half a century. I understand that in 1994 the Chinese Congress decided to go ahead with the project. Most of the cost will be borne by China, but some outside financing will be sought from organizations like the World Bank. The relocation work has nearly been completed.

I gave several talks at the planning office for Wei Tingcheng, the main proponent of the project. Before I left Wuhan I was appointed as a consultant to the dam project, a position I gladly accepted.

The latest news from China is that Jiang Jenming has taken over the top position in the government. Recent developments indicate to me that China is leaning toward becoming an industrial nation. China has yet to resolve its population problems and issues of corruption in the government. The latter especially resulted in the downfall of the Guomingdang during the years I worked in China. The United States might try to view China from a broader perspective, rather than dwelling endlessly on human rights issues.

*Zhao Zukang (second from left), author's former boss and
Chinese highway commissioner, 1979, Beijing*

*Reunion with Qinghai-Tibet Highway senior engineer, 1979
(left to right) Lieu Cheng Xian, author, and Li Chang*

Author at site of the Three Gorge Dam and rock-drilling cores,
which are among the largest in the world

– Chapter Twenty-Three
Career in Crest

In 1980 I was elected president of the American Society of Civil Engineers of Colorado. First on my agenda was to utilize the voices of our nineteen hundred members as a political voice. We were interested in influencing legislation that would affect the practice of engineering in Colorado, including issues of professional registration, energy, water resources, and the environment.

Since I had been president of the three major engineering organizations in Colorado, I raised the issue of unifying the engineers. I was not advocating the merger or consolidation of the societies, since each has its own history and areas of service. However, there was much duplication of efforts, which wasted time and resources.

My first step was to publish a joint newsletter, with all three secretaries sharing the fiscal responsibility. The first issue appeared in January 1983. Eventually, cooperation among the societies extended to the sharing of offices, computers, secretarial services, and employment listings. Colorado was somewhat behind other states in initiating intersociety efforts.

I received the Orley O. Phillips award from my fellow engineers in 1982 for outstanding contributions to the engineering profession in Colorado. In January 1981 I was elected vice president of the American Consulting Engineers Council. During my two-year term of office, I represented seven western states on the board of the national council.

The American Consulting Engineers Council is the largest national organization of engineers engaged in consulting as a private practice. The council consists of nearly six thousand member firms that represent around two hundred thousand employees. The member firms range in size from a single experienced registered professional to corporations employing thousands of professionals with diverse expertise.

In 1982 I was elevated to senior vice president of the council, in charge of the International Affairs Committee. I reported on our ongoing involvement with various projects in China. Although I was disappointed that our work with the Chinese did not show any immediate result, I believe the council plays a major role in providing information, making contacts, and exploring the Chinese market for member firms.

I remained opposed to the national U.S. trend toward quota systems, or "set asides," for various privileged groups. Just as we have only one class of citizenship in the United States, I firmly believe all citizens should be treated equally. Business competition should be based on merit alone.

For many centuries China had an examination system for the selection and placement of government officials. In the traditional system, "Jinshi" represented the highest accomplishment, equivalent to the U.S. doctoral degree. In China, any scholar who obtained Jinshi was honored by everyone and could obtain audiences with the emperor. Jinshi were usually appointed to top positions in the regime. Scholars sometimes devoted their entire lives to trying to achieve this honor. This concept is so deeply ingrained in the minds of the Chinese that even today, both in China and in Taiwan, the common conception is that engineers with Ph.D.s should get the highest compensation. The concept is especially prevalent in academic circles. In the United States, the Ph.D. is almost like a license to teach.

In the workaday world of engineering, experience commands much more respect than academic degrees. In the American Consulting Engineers Council, experience is the key determinant of position and salary. Yet, although I had fifty years' experience as a geotechnical engineer, some of my fellow engineers had still considered my knowledge inferior to that of someone with a doctorate. So when Colorado State University (CSU) awarded me the honorary degree Doctor of Science on August 19, 1980, my delight and excitement were beyond description. I had attained something I had always wanted, and the United States had given me much more than I thought I deserved. The citation from CSU read in part:

> Fu Hua Chen, distinguished engineer, educator, and adminis-trator, your technical contributions and service to the profession have earned you a high rank among your peers. Your career has had a direct impact on engineering and technology in the area of foundations and soils, and you have had a great deal of influence on the development of young engineers. . . .
> Fu Hua Chen, your outstanding record as a consultant as well as an academician makes you truly a credit to the profession of civil engineering, and it is my distinct pleasure today, under the authority vested in me by the governing board of Colorado State

University, to confer on you the honorary degree Doctor of Science, with all the rights, the honors, and privileges appertaining thereto.

I spoke at the commencement exercises. Around 1,275 seniors and graduate students received degrees at that summer commencement. With families and guests, more than 3,000 people attended, making it the largest group I had ever addressed. In my speech I stressed the importance of private enterprise. I got a standing ovation for these words: "As a minority, I believe in equal opportunity and fair competition. The quota system of minority participation not only is not fair to the industries and educational institutions but will deprive minorities of their dignity and will, in the long run, hurt the cause. I do not want people to address me by my color or creed, but I will be proud to be addressed as 'my fellow American.'"

When the old Glover building at Colorado State University was renovated and became the new engineering building, the school named a soil laboratory after me. I felt a special delight in seeing my portrait hanging on the wall next to all the testing equipment. In 1983 Colorado State University gave me the personal service award at the Dean's Council dinner before 140 alumni and friends. Lionel Baldwin, dean of engineering, said,

> Fu Hua has worked with the university for many years in several capacities — as an affiliate faculty member, as a practicing engineer, as an active lobbyist on behalf of CSU engineering with the Colorado legislature. He has done much to improve the image of the engineering profession in the state of Colorado. Through his leadership in professional societies, he has assisted first-hand in the learning experiences of Colorado State University engineering students, and he has spent countless hours during the last few years to improve the level of support directed from the Colorado legislature to engineering higher education.

I delivered the first annual CSU geotechnical engineering distinguished lecture on May 14, 1985. The series has continued in subsequent years. My lecture was followed by a lecture by Professor Ralph Peck, for whom I have great respect and admiration. I had translated his book into Chinese when I was in Shanghai.

On September 25, 1986, Governor Lamm honored six Colorado scientists, including me, with the first Governor's Award for Science and Technology. Lamm remarked, "It is a source of great pride to Colorado to have such distinguished scientists in our population. These people show us continuously

Fu Hua Chen Geotechnical Laboratory at Colorado State University
(Professor John Nelson and author)

that creativity can bring very tangible results that improve all our lives immensely. Not only are their works worthwhile, but they give us promise for the future by their example. They have honored us as much as we can honor them."

The citation stated "Chen, 70, of Denver, a Coloradan for thirty years, helped design the caisson building foundation system extensively used along the front range. He was chief engineer for the Burma Road and the Koknor-Tibet Highway and authored *Foundations on Expansive Soils,* the only book on the subject. His firm has been involved in more than 35,000 projects throughout the Rocky Mountain Region."

On June 27, 1986, I was elected to the grade of Honorary Member of the American Society of Civil Engineers (ASCE) by the society's board of directors. The group recognized my fifty-year contribution to civil works, study of expansive soils, publications, and activities in professional groups and education. Election to this position requires that the engineer has demonstrated eminence in some branch of engineering or in the related arts and sciences. Since 1882 the ASCE has awarded only 104 honorary memberships, including one to Herbert Hoover. I believe I was the first from Colorado to receive the award.

Earlier, my efforts to unify the engineers had been recognized by the Colorado Engineering Council, an amalgam of all the major engineering organizations in the state. The council awarded me its greatest honor, the gold medal award, which is reserved for those the council considers exemplary in service to the profession and society. I will treasure the medal always; it was the first award I ever received that was struck in real gold.

– Chapter Twenty-Four
Stepping Down

After twenty-three years as president of my own company, I decided to retire. I had seen business executives like myself hang onto the companies they had founded to their last breath. As a result, perhaps as a result of poor management in their later years or of the ineffectiveness of their heirs, the companies deteriorated. One striking example of this was Dr. Wang An, the founder of Wang Laboratories, the electronics giant. Wang An wanted to continue to control his empire and arranged for his son to take it over. Within a few years, the name Wang Laboratories had gradually faded in the business world.

At the annual company party in 1983, I gave a talk to more than one hundred employees and guests. I recognized those who had helped build up my business during the twenty-three years. We had grown from a one-man operation to a company with more than two hundred employees and annual billings exceeding $10 million. The *Engineering News Record* listed us among the one hundred largest engineering companies in the United States.

I concluded by saying, "It is not possible to tell you what has transpired in the last twenty-three years in thirty minutes. There were moments of elation, moments of gratification, moments when I stopped to think about how a foreigner was able to achieve his dream in this land of opportunity. On the other hand, there were moments when I wondered about the merit of our legal system, moments when I doubted whether all this is worthwhile. I wish I had the faith of my wife, Edna, in God, which enables her to face everything with confidence." My closing remark was "After all that has been said and done, it has been an exciting twenty-three years. As the Chinese say, 'All feasts must come to an end.' As the Americans say, 'It is time that we got a new quarterback.'" With those remarks I stepped down as president, and Dick Hepworth took over as the new president of Chen and Associates.

I had plenty of other activities planned. First, I revised my book *Foundations on Expansive Soils*, which had originally been published in 1973. It was still the only book in the world on the subject. When I finished updating it with new information, the volume had grown to twice the size of the original.

Since 1965 an international organization for research on expansive soils had existed. The first conference was held at Texas A&M University. I presented a paper, and thus began my participation in international activities related to expansive soils. Subsequently, conferences were held in Haifa, Israel; Denver, Colorado; Sidney, Australia; New Delhi, India; and Dallas, Texas. I attended and was active at all of the conferences, and I enjoyed the opportunity to meet the outstanding scholars in every country and broaden my knowledge.

I always like to relate the story of the blind men and the elephant in my speeches, because I believe scientists and engineers are groping toward narrow and limited areas of their specialties and have not been able to see the entire picture but only see one side. A lot of our effort could be avoided if we understood the structure of the universe. That might be the day when science and religion finally meet and merge. Chinese philosophers talked and wrote about such concepts even before the turn of the twentieth century.

Engineers of Chinese descent in Colorado organized the Colorado Chinese Society of Science and Engineering in 1984. The first meeting was held in September of that year, and I was invited to speak. The goal of the organization was to promote the advance of science and engineering in the Chinese American community. I have high hopes for the society, because this is the first time I have witnessed communication and unity among Chinese American engineers, many of whom were immigrants like myself. For many years, those who favored Taiwan and those who wished to work with China pulled in different directions. To bring them together, we needed a nonpolitical organization. The society has flourished in recent years and is respected among the engineering societies.

Diverse opinions flourish in the Colorado engineering community. Most believe we should maintain our technical excellence and leave politics to politicians. Unfortunately, the politicians are well aware of this feeling and still take advantage of engineers by passing laws like those that deal with liability and warranty without the input of the engineering community.

I firmly believe, as before, that the various engineering societies should have a strong voice in the U.S. Congress, as well as in state legislatures, and that Chinese American engineers should run for political office. Other minorities have more representation in the political process. For the engineering profession, getting into the political arena can be as important as the advancement of science and technology.

The Chinese American community in the United States has undergone profound changes in the past fifty years. Instead of being dominated by immigrants from Guangzhou, it now includes Chinese students from other areas of China seeking educations in U.S. universities and descendants of Chinese immigrants from all over the country. The Chinese laundry has become a thing of the past, but the restaurant business still flourishes in most U.S. cities, large and small. Chinese American students excel in all areas of scholarly endeavor. Today, few universities or colleges have no professors of Chinese descent. We have also become important contributors to government work, at both the federal and state levels.

The Chinese American community in Denver gave my wife and me a surprise eightieth birthday party, with more than a hundred people in attendance. They presented me with a huge plaque, which I treasure. I gave a long talk in Chinese on the events of failure in my life; I feel those are more exciting and more meaningful than success stories.

With time on my hands, I returned to my lifelong hobby of stamp collecting. Although I had a good collection of stamps from Germany, Belgium, France, Italy, the Netherlands, Norway, and many other countries, I now chose to specialize in Chinese stamps, since I was familiar with that country's history, political divisions, and people. Perhaps most important, Chinese stamps still represent China's century of humiliation, as many were issued by foreign governments out of their various territorial concessions.

Before my decision to step down as president of the company, and in view of the successful operation of Chen and Associates, we had been approached by several nationally known engineering consulting companies about a buyout or a merger. These business practices have become common in the engineering world and in U.S. business in general. The larger a company grows, the more difficult it becomes to maintain an adequate cash flow. Financially, Chen and Associates had the backing of the First Interstate Bank, but it had become obvious that we either had to accept our size as it was or be absorbed by the so-called industry giants.

I had to weigh the future of the company for some time. In other companies, I had seen cases of successful operations being forced into bankruptcy because of cash-flow problems. Also, my age was catching up with me, and I simply could no longer endure the extensive international travel that was necessary for staying on top of the business. After long meditation and many discussions with the major stockholders, we decided to consider the various offers.

Most of the offers we received involved the exchange of stock and a firm commitment to allow employees to stay with the company for an extended period. I found these terms unacceptable. Then a British corporation came forward, Huntington International Holdings (HIH). We had lengthy talks, and finally, in March 1988, we reached an agreement. HIH would pay cash for our stock with no strings attached.

When I thought back twenty-seven years to the days when everyone warned me against starting my own company, I would never have dreamed that my stock would now sell for a thousand times my original investment. By that time I had divested myself of the initial stock of five thousand shares, which were now held by seventeen employees. Key people now received the lion's share of the stock. I had no complaints. On April 30, 1988, Huntington International Holdings officially took over Chen and Associates. For the time being, I remained chairman of the board.

On that day, all seventeen stockholders were present, together with our faithful attorney, William Caskins, and a representative of the First Interstate Bank. HIH representatives handed each stockholder a check, with no employment commitment. Thus, Chen and Associates officially ceased to exist. It broke my heart to see our sign taken down at the front of the building. HIH wanted to retain the name "Chen" in the logo as part of the agreement.

In the years since I stepped down as president of Chen and Associates, a great deal of change has occurred in the company, although it remains one of the leading consulting firms in the United States. The company has changed from a close-knit organization to a typical faceless corporation where performance is measured by productivity, time cards are closely scrutinized, and overtime must be authorized. In the old days, I could always depend on my employees to come in on Saturday or Sunday when necessary. We did not have to check on the amount of time our people spent on any project. Also, we had the reputation of never having lost a client. In the past four years the company has gradually lost its personal touch and has followed the apparently inevitable path of U.S. business.

Nearly all of the former employees of Chen and Associates have left the new organization. I have found that HIH is running the company in the way of any retail business, with focus on the bottom line. Profit is the main consideration. Of my former staff of two hundred employees, only four remain in the company. I am pleased to note that all of those who left the company have excellent employment elsewhere. Some have started their own companies and are doing very well. Others are working for my former competitors at higher pay. I have yet to know of anyone who is unemployed.

We have an annual reunion party, and more than one hundred people came to the last one. I gave a speech at the party, thanking them for the hard work they had rendered in the past. In particular I thanked Dick Hepworth, who was responsible for the growth of the company during its last five years.

I am now eighty-four years old, and as the Chinese say, it is rare that our years exceed seventy. All of my life I have enjoyed excellent health, even through the physical hardships I have endured. Now my health is deteriorating. In the last year, I suffered a heart attack and required a pacemaker, then I had a stroke and can no longer drive. I have a severe case of emphysema, no doubt as a result of heavy smoking in my younger days.

Looking back, I have enjoyed good days, but, of course, one always wishes for a chance to relive the past and change it. I am grateful to my family. Edna helped my business when I needed help the most, and she always gave me courage when I was depressed. My daughter, Dorothy, has led an outstanding life. She earned a Ph.D. from the University of Texas while she held a full-time job and took care of her children. She is an immunologist in charge of a laboratory at the University of Texas Hospital and gives lectures all over the United States. Her daughter, Jean, went to school at Wellesley College and has an advanced degree from the University of Michigan, where I got my engineering degree.

I was in Dallas attending the Seventh International Conference on Expansive Soils when my granddaughter, Jean, married a young man from China. It was a memorable event, as Colorado State University also gave me a reception to celebrate my birthday. I was pleased to know that my name was familiar to engineers from France, Israel, Australia, India, and, of course, China.

Dorothy's son, Lester, graduated first in his class at Rice University with a degree in mechanical engineering and had his choice of scholarships from MIT and Stanford. He decided to go to California, because it is more fun than Boston. I am amazed to see that our second generation is smarter than we were. They did not have to study as hard and yet have accomplished a great deal.

My son, Tyrone, worked for me briefly. I have to say that I treated him roughly; I did not want my employees to think he was getting special privileges. Tyrone got his master's degree from the University of Colorado and has a good job at Hughes in Los Angeles. He married a medical doctor from China and is enjoying life in California.

My younger daughter, Yvonne, had nothing to do with science and engineering; she chose to study library management. After many years of hard work, she is now director of the Seattle Public Central Library. Her son is enrolled at Harvard.

I am grateful that our children maintain the traditional respect and devotion to their parents. When I was sick and in the hospital, they came to take care of me. Considering that I did not spend much time in bringing them up, they did more for me than I deserve.

All through my childhood and during my adult years, I heard the word *China* linked with poverty, flood, famine, and misery. Missionaries in China boasted about their charity work. The United States sent food and clothing to China for humanitarian causes. We saw advertisements for saving a child in China by sending a dollar. In the early decades of the twentieth century, the common perception in the United States seemed to be that the warlords in China were feasting on the poor and gaining enormous wealth. The very word *China* signified civil war, death, and injustice. The darkest place on earth was China. We were ashamed of being Chinese.

The century of humiliation only got worse. At the height of the Sino-Japanese War, westerners viewed the episode as a local affair and glorified the Japanese as the power in Asia. China was viewed as having to survive on U.S. charity, without which the country would have vanished.

When the Communists took over China, the Western world viewed the event as another falling domino that would not improve the miserable conditions in China. To the contrary, China refused all Western aid, including that from the Soviet Union. Mao believed Chinese problems must be resolved by the Chinese alone.

The rebuilding of China started in 1950. Today, China stands as an independent nation. Dams were built to curb flooding; railroads were built to balance the distribution of the population; textile mills were built to clothe the population; farm production was increased to feed the hungry. Certainly, major setbacks occurred, especially during the cultural revolution, yet the transformation of China from a country seeking handouts to a self-sufficient, industrialized nation is remarkable. Even following the major earthquake at Tangshan, China refused international aid. I do hope the present regime will continue its effort to curb corruption and control population growth. In my remaining years, I hope to see a strong and prosperous China.

U.S. politicians firmly believe democracy is the only correct form of government and that all nations should follow the American way. Democracy has performed well in the United States and in some European countries, and

Stockholders (left to right): Marc Pardi, Dick Hepworth, Jeff Johnson, Narender Kumar, Ken Temme, Dave Adams, Dave Jubenville, Roger Barker, Fu Hua Chen, Art Greengard, Harold Hollingsworth, Rich Tocher, Hans Froeschale, Dave Herman, Jim Nordquist, Ralph Mock, and Don Bressler.

Americans would not consider any other form of government. On the other hand, in China, where more than a billion people must be governed, most citizens have no idea about democracy. As the Chinese saying goes, "The primary law of heaven is to feed the people."

I do not favor the Communist form of government, but I also believe the movement for human rights must be generated from within and cannot be implemented by the threat of other nations. Could American-style democracy have any success in Asian countries? I have serious doubts that it could.

When I awake in the middle of the night, I often ask myself, "Is life worth living?" When I pass away, my life will have been like striking a match.

Although the flame appears rosy, it lasts only a few seconds and is followed by darkness, and nothing can rekindle the match.

My wife's answer would be, "Turn to God, and you will find peace and happiness." That has not worked for me. I have yet to find a religion I can put that kind of faith in.

I have gone through the vicissitudes of life, which have been interesting and exhilarating. Certainly, it has not been dull. Now I enjoy my retirement in an afterglow, like the sunset.

– Epilogue

Since retiring from Chen and Associates, I have given lectures, seminars, and specialty courses in expansive soils at the University of Colorado at Denver. In 1991, I was appointed professor adjunct of civil engineering and established the Expansive Soil Research Center (ESRC) in addition to teaching courses in behavior and control of expansive soils. In further developing the ESRC, I donated and catalogued my personal library of international literature and photographs to provide a reference source for geotechnical engineers. I have just been informed that in recognition of my contributions to CU–Denver and the engineering profession, I have been awarded the 1996 University of Colorado at Denver Alumni Appreciation Award.

Map of West China: Key Roads and Cities

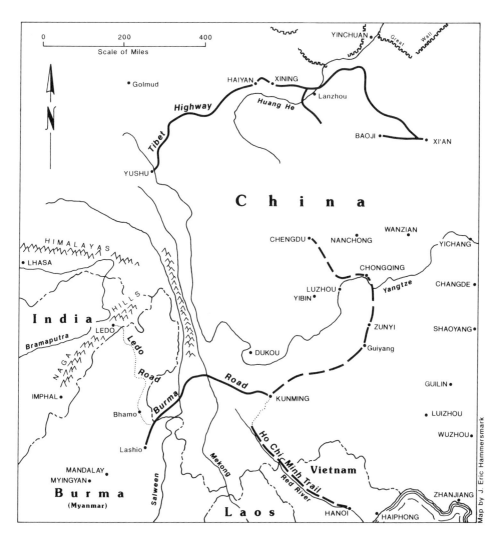

The heavy solid lines denote roads on which the author participated in the construction. Roads shown by heavy dashed lines were traveled or reconnaissanced by the author. Dotted lines represent a connecting road or trail. Cities shown in ALL CAPS were visited by the author. Geographical locations are indicated in the pinyin system.

Translation and Characters

For years, the Chinese alphabet and romanized spelling were based on the *National Language Dictionary,* and it was the most widely recognized system used. After 1949 the new system of translation of Chinese characters, known as "pinyin," was adopted by the Chinese and others in an effort to spell words more as they sound when spoken with standard Beijing pronunciation. Some words may not be familiar. To help them become more so, this table includes the new system, the old system, and the Chinese characters. Throughout the entire book, the pinyin system has been used.

Modern Transliteration	Old Transliteration or English	Chinese Character
Ah Sheng	Au Shen	阿　勝
Ai Feng		靄　峰
Ai Di		哀　常
An Lushan	An Lo Shan	安　祿　山
Animaching Peak		阿尼瑪鄉峰
Aulinghu	Au Ling Lake	鄂　陵　湖
Badicheng	Po Ti Chen	白　帝　城
Bamo	Pa Mo	八　莫
Bang	Ban	幫
Baoji	Pao Chi	寶　雞
Baoshan	Pao Shan	寶　山
Beijing	Peking	北　京
Chalinghu	Chaling Lake	扎　陵　湖
Changan		長　安
Changsha	Chang Sha	長　沙

Chelaping		查	拉	坪
Chen Mu		陳		模
Chen Sheng	Chen Shen	陳		勝
Chengdu	Cheng Dao	成		都
Chongqing	Chungking	重		慶
Chuxe	Chuchiehsu	竹	節	寺
Ci Xi	Tzu Hsi	慈		禧
Cultural revolution		文 化 革 命		
Dali	Ta Li	大		理
Damqu River		曲	當	河
Deng Ai		鄧		艾
Devil's Abode		閻	王	溝
Dinghai	Ting Hai	定		海
Dujiangyan		都	江	堰
Emei	Mount Omei	峨		眉
Evil Mist		瘴		氣
Fuzhou	Fukien	福		州
Fu Tan		复		旦
Gang of Four		四	人	幫
Gansu	Kansu	甘		肅
Geladandong		各 拉 丹 冬		
Grezhouba		葛	州	壩
Guangzhou	Canton	廣		州
Guiyang	Kwei Yang	貴		陽

Guizhou	Kwei Chow	貴		州
Guomindang	Kuomintang	國	民	黨
Han		漢		
Hangzhou	Hang Chow	杭		州
Hankou	Hankow	漢		口
Han Yu		韓		愈
Hengduan		橫		斷
Hu Shi	Hu Shih	胡		適
Huangguoshu		黃	果	樹
Huanghe	Hwang Ho	黃		河
Huangheyan		黃	河	沿
Huangpu		黃		浦
Huangyuan		湟		源
Hui		回		
Huihai	Huai hai	淮		海
Hunan		湖		南
Jialing	Chailin	嘉		陵
Jiang Jenming		江	澤	民
Jiang Jieshi	Chiang Kaishek	蔣	介	石
Jiang Xing Kuo		蔣	經	國
Jingjiao	Nestorianism	景		教
Jinshi		進		士
Ju Ru	Marchny Meadow	沮	汝	地
Kaiyuan	Kei Yuan	開		遠
Kang Xi	Kang Hsi	康		熙

Kazakh		哈	薩	克
Kunlun		崑		崙
Kunming	Kumming	昆		明
Lachu	Lachio	臘		戎
Lancang	MeKong	瀾		滄
Lanzhou	Lan Chow	蘭		州
Leshan	Loshan	樂		山
Lhasa		拉		薩
Li Bai	Li Bai	李		白
Li Bing		李		冰
Li Changyuan	Li Chang Yuan	李	昌	源
Lijiang		漓		江
Lin Sen		林		森
Lin Zexu	Lin Tse Hsu	林	則	徐
Litchi		荔		枝
Liu Chengxian	Liu Cheng Hsien	劉	承	先
Liu Rusong	Liu Ju Song	劉	如	松
Liu Zhou	Liu Chow	柳		州
Lu	Lun	瀘		
Lugouqiao		蘆	溝	橋
Manzhou	Manchu	滿		州
Mao Yisen	Mao I Sen	茅	以	升
Mao Zedong	Mao Tze Tung	毛	澤	東
Matu		瑪		多

Mawangdui		馬 王 堆
Mawei		馬 嵬
Mejiang	Mekong	閩江，湄江
Mengtzu	Mang Tsu	蒙 自
Ming		明
Ming Huang		明 皇
Mingshan		岷 山
Na Keng		那 坑
Nanjing	Nanking	南 京
Nanjinguan		南 津 關
Ngolog		聶 果 洛
Nei Baizheng	Nei Pe Cheng	聶 百 徵
Ningbo	Ning Po	寧 波
Nujiang	Salween	怒 江
Pai Chuyi	Bai Ju Yi	白 居 易
Pan Zufeng		潘 祖 芬
Payekala		巴 顏 克 拉
Puxian		普 西
Qian Long		乾 隆
Qin	Chin	秦
Qing	Ching	清
Qingdao		青 島
Qinghai	Kokonor	青 海
Qinghai-Tibet Highway		青 藏 公 路
Qutang		瞿 塘

Salween River Bridge		惠 通 橋
Sandouping		三 斗 坪
Sangzanganbu		松 贊 干 布
Shamen	Amoy	廈 門
Shantou	Swato	汕 頭
Sichuan	Szechuan	四 川
Six Fold Mountain		六 盤 山
Song	Soong	宋
Song May-ling		宋 美 齡
Star Cluster Sea		星 宿 海
Stone Forest		石 林
Sun Yixian	Sun Yat Sen	孫 逸 仙
Sun Zhong Shan	Sun Chung Shu	孫 中 山
Suzhou	Suchow	蘇 州
Taer	Taer Monastery	塔 爾 市
Tai Zhong		太 宗
Taihangshan		太 行 山
Tang		唐
Tang En Bo		湯 恩 伯
Tangchang		宕 昌
Tanggulia		東 鄂 拉 山
Tangtianhe		通 天 河
Tengchong		騰 沖
Three Gorge Dam		三 峽 大 壩
Tienjin	Tientsin	天 津
Touching Heaven Mountain		摩 天 嶺
Tufan		吐 蕃

University of Nankai		南 開 大 學
Wang An		王　　安
Wanting		宛　　汀
Wei Tingcheng		魏 延 玼
Wen Cheng		文　　成
White Dragon River		白 龍 江
Wu Junwei		伍 俊 威
Wudu		武　　都
Wuhan		武　　漢
Wushan		巫　　山
Wuxia		巫　　峽
Xewu	Hsieh Wu	歇　　武
Xiaguan	Hsia Kwan	下　　關
Xian	Sian	西　　安
Xianyang	Hsien Yang	咸　　陽
Xichang	Sichang	西　　昌
Xikang	Sickang	西　　康
Xikang-Tibet Highway		康 藏 公 路
Xiling		西　　陵
Xining	Hsi-Ning	西　　寧
Xinjiang	Sinkiang	新　　疆
Xu Baoshan	Hsu Bao Shan	徐 寶 山
Yalujiang	Yalung Kiang	鴨 綠 江
Yang Kuefei		楊 貴 妃

Yang Zhenning		楊	振	寧
Yangjing	Yenching	燕		京
Yangshu		陽		朔
Yangzhou	Yang Chow	揚		州
Yangziejiang	Yangtze River	揚	子	江
Yantai	Chefoo	煙		台
Yarlungzangbo		雅 魯 藏 布		
Yematan	Yeh Ma Tan	野	馬	灘
Yibin	I Pin	宜		賓
Yichang	I Chang	宜		昌
Yong He Palace		雍	和	宮
Yong Zheng		雍		政
Yu Feipeng		俞	飛	鵬
Yuan Shikai		袁	世	凱
Yunnan	Yunan	雲		南
Yushu		玉		樹
Zanba		糌		巴
Zhao Zi Yang		趙	紫	陽
Zhao Zukang	T. K. Chou	趙	祖	康
Zheng Yangfu	Tseng Yang Fu	曾	養	甫
Zhu Guoxi		朱	國	洗
Ziu Zhou	Liu Chow	柳		州

Index

Acupuncturist, encounter with, 45–46
Adams, Dave: photo of, 171
Agent Orange, effects of, 71
Aggregates, preparing, 85 (photo)
Ai Di, Emperor: homosexuality of, 14
Ai Feng, 153
Al-Anbar, Nasser: meeting with, 141–142
Alpha Lambda fraternity, starting, 18
American Consulate (Hong Kong), interview at, 114–115
American Consulting Engineers Council, 136, 137; memorandum of understanding with, 151, 152; serving on, 160, 161
American Society of Civil Engineers (ASCE), 136; as Colorado president of, 160; honorary membership to, 163–164
Ancestors, respect for, 44
Anderson, Al, 126
Animachang Peak, 99
An Lushan, 31
Anti-Chinese movement, growth of, 16
Armed guards, hiring, 50, 52
"Army From Heaven," legend of, 46
ASCE. *See* American Society of Civil Engineers
ASFE. *See* Association of Soil and Foundation Engineering
Asphalt: acquiring, 82; hand application of, 86 (photo)
Asphalt Institute, 83; manual by, 82
Asphalt Pavement Construction (manual), 82
Association of Soil and Foundation Engineering (ASFE), founding of, 130–131
Atabrine, smuggling, 5
Attorneys, problems with, 14, 137

Baldwin, Lionel, 162
Bandits, encountering, 26, 50, 52
Bang. *See* Secret societies
Baoshan: road building at, 80; traveling to, 90
Barges: photo of, 107; river crossing by, 63
Barker, Roger: photo of, 171
Bedbugs, 35–36, 40
Beijing: high school in, 13; Japanese emperor at, 60; visit to, 146–148, 150, 152–153
Bentonite, problems with, 127
Berlin Olympics, attending, 19
Bicycles, problems with, 26
Black market, 65, 73
Blasting, problems with, 53–54

Bomb shelter (Chongqing), tragedy at, 59
Bottle-Shaped Pass, ambush at, 32
Bowen, Bill, 126, 131
Bressler, Don: photo of, 171
Bridge building, 3–4, 77
British Civil Service, 114
British Concession, 10, 11
Buddha Puxian, statue of, 75
Bureaucracy, 120; dealing with, 61, 69, 73, 138–139
Bureau of Reclamation, Hoover Dam and, 152
Burma Road, xii, 2, 103, 104, 109, 114, 148, 163; assignment to, 78; constructing, xi, 1, 4, 5, 61, 83; Japanese on, 88, 89, 91–92, 93; length of, 79; map of, 84; monument on, 92; photo of, 84; public story of, 87; traffic on, 6 (photo), 7 (photo), 82, 83
Bynum, Mr., 145

CACI. *See* Colorado Association of Commerce and Industry
Cannibalism, 93–94
Capital Construction Division (State Economic Commission), American Consulting Engineers Council and, 152
Carter, Jimmy, 145
Caskins, William, 168
Cement and soil, mixing, 18, 22, 23, 24, 137
Central Mining Company, equipment from, 144
Central Pacific Railroad, 16, 52
Changan, 21, 31
Changjiang (Long River), 103
Changsha, visit to, 154
Chen, Dorothy, 121, 123; birth of, 87; career of, 169; photo of, 91
Chen, Fu Hua: photo of, 76, 91, 106, 116, 158, 159, 163, 171
Chen, Tyrone, 121, 146; birth of, 108; career of, 169
Chen, Yvonne, 121, 146, 152; birth of, 108; career of, 170
Chen and Associates: considering buyout/merger, 167–168; growth of, 126, 129–130, 131, 134, 165; Kodak contract for, 126–127; lawsuit against, 131–132; as minority enterprise, 139; national recognition for, 135; offices of, 128 (photo), 129 (photo), 130 (photo), 131; overseas

185